Boundary Control of PDEs

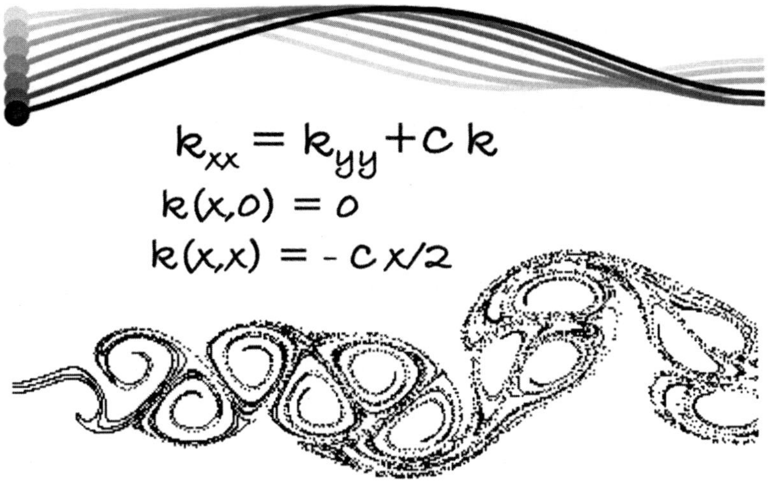

Advances in Design and Control

SIAM's Advances in Design and Control series consists of texts and monographs dealing with all areas of design and control and their applications. Topics of interest include shape optimization, multidisciplinary design, trajectory optimization, feedback, and optimal control. The series focuses on the mathematical and computational aspects of engineering design and control that are usable in a wide variety of scientific and engineering disciplines.

Series Volumes

Speyer, Jason L., and Jacobson, David H., *Primer on Optimal Control Theory*
Betts, John T., *Practical Methods for Optimal Control and Estimation Using Nonlinear Programming, Second Edition*
Shima, Tal and Rasmussen, Steven, eds., *UAV Cooperative Decision and Control: Challenges and Practical Approaches*
Speyer, Jason L. and Chung, Walter H., *Stochastic Processes, Estimation, and Control*
Krstic, Miroslav and Smyshlyaev, Andrey, *Boundary Control of PDEs: A Course on Backstepping Designs*
Ito, Kazufumi and Kunisch, Karl, *Lagrange Multiplier Approach to Variational Problems and Applications*
Xue, Dingyü, Chen, YangQuan, and Atherton, Derek P., *Linear Feedback Control: Analysis and Design with MATLAB*
Hanson, Floyd B., *Applied Stochastic Processes and Control for Jump-Diffusions: Modeling, Analysis, and Computation*
Michiels, Wim and Niculescu, Silviu-Iulian, *Stability and Stabilization of Time-Delay Systems: An Eigenvalue-Based Approach*
Ioannou, Petros and Fidan, Barış, *Adaptive Control Tutorial*
Bhaya, Amit and Kaszkurewicz, Eugenius, *Control Perspectives on Numerical Algorithms and Matrix Problems*
Robinett III, Rush D., Wilson, David G., Eisler, G. Richard, and Hurtado, John E., *Applied Dynamic Programming for Optimization of Dynamical Systems*
Huang, J., *Nonlinear Output Regulation: Theory and Applications*
Haslinger, J. and Mäkinen, R. A. E., *Introduction to Shape Optimization: Theory, Approximation, and Computation*
Antoulas, Athanasios C., *Approximation of Large-Scale Dynamical Systems*
Gunzburger, Max D., *Perspectives in Flow Control and Optimization*
Delfour, M. C. and Zolésio, J.-P., *Shapes and Geometries: Analysis, Differential Calculus, and Optimization*
Betts, John T., *Practical Methods for Optimal Control Using Nonlinear Programming*
El Ghaoui, Laurent and Niculescu, Silviu-Iulian, eds., *Advances in Linear Matrix Inequality Methods in Control*
Helton, J. William and James, Matthew R., *Extending* H^∞ *Control to Nonlinear Systems: Control of Nonlinear Systems to Achieve Performance Objectives*

Boundary Control of PDEs

A Course on Backstepping Designs

$$k_{xx} = k_{yy} + c\,k$$
$$k(x,0) = 0$$
$$k(x,x) = -cx/2$$

Miroslav Krstic
University of California, San Diego
La Jolla, California

Andrey Smyshlyaev
University of California, San Diego
La Jolla, California

siam.

Society for Industrial and Applied Mathematics
Philadelphia

10 9 8 7 6 5 4 3 2

Library of Congress Cataloging-in-Publication Data

Krstic, Miroslav.
Boundary control of PDEs : a course on backstepping designs / Miroslav Krstic, Andrey Smyshlyaev.
p. cm. – (Advances in design and control ; 16)
Includes bibliographical references and index.
ISBN 978-0-898716-50-4
1. Control theory. 2. Boundary layer. 3. Differential equations, Partial. I. Smyshlyaev, Andrey. II. Title.
QA402.3.K738 2008
515'.353–dc22 2008006666

Contents

Preface

The method of "integrator backstepping" emerged around 1990 as a robust version of feedback linearization for nonlinear systems with uncertainties. Backstepping was particularly inspired by situations in which a plant nonlinearity, and the control input that needs to compensate for the effects of the nonlinearity, are in different equations. An example is the system

$$\begin{aligned}\dot{x}_1 &= x_2 + x_1^3 d(t),\\ \dot{x}_2 &= x_3,\\ \dot{x}_3 &= u\,,\end{aligned}$$

where $x = (x_1, x_2, x_3)$ is the system state, u is the control input, and $d(t)$ is an unknown time-varying external disturbance. Note that for $d(t) \equiv 1$ this system is open-loop unstable (with a finite escape time instability). Because backstepping has the ability to cope with not only control synthesis challenges of this type but also much broader classes of systems and problems (such as unmeasured states, unknown parameters, zero dynamics, stochastic disturbances, and systems that are neither feedback linearizable nor even completely controllable), it has remained the most popular method of nonlinear control since the early 1990s.

Around 2000 we initiated an effort to extend backstepping to partial differential equations (PDEs) in the context of boundary control. Indeed, on an intuitive level, backstepping and boundary control are the perfect fit of method and problem. As the simplest example, consider a flexible beam with a destabilizing feedback force acting on the free end (such as the destabilizing van der Waals force acting on the tip of the cantilever of an atomic force microscope (AFM) operating with large displacements), where the control is applied through boundary actuation on the opposite end of the beam (such as with the piezo actuator at the base of the cantilever in the AFM). This is a prototypical situation in which the input is "far" from the source of instability and the control action has to be propagated through dynamics—the setting for which backstepping was developed.

Backstepping has proved to be a remarkably elegant method for designing controllers for PDE systems. Unlike approaches that require the solution of operator Riccati equations, backstepping yields control gain formulas which can be evaluated using symbolic computation and, in some cases, can even be given explicitly. In addition, backstepping achieves stabilization of unstable PDEs in a physically appealing way, that is, the destabilizing terms are eliminated through a change of variable of the PDE and boundary feedback. Other methods for control of PDEs require extensive training in PDEs and functional analysis.

Backstepping, on the other hand, requires little background beyond calculus for users to understand the design and the stability analysis.

This book is designed to be used in a one semester course on backstepping techniques for boundary control of PDEs. In Fall 2005 we offered such a course at the University of California, San Diego. The course attracted a large group of postgraduate, graduate, and advanced undergraduate students. Due to the diversity of backgrounds of the students in the class, we developed the course in a broad way so that students could exercise their intuition no matter their backgrounds, whether in fluids, flexible structures, heat transfer, or control engineering. The course was a success and at the end of the quarter two of the students, Matthew Graham and Charles Kinney, surprised us with a gift of a typed version of the notes that they took in the course. We decided to turn these notes into a textbook, with Matt's and Charles' notes as a starting point, and with the homework sets used during the course as a basis for the exercise sections in the textbook.

We have kept the book short and as close as possible to the original course so that the material can be covered in one semester or quarter. Although short, the book covers a very broad set of topics, including most major classes of PDEs. We present the development of backstepping controllers for parabolic PDEs, hyperbolic PDEs, beam models, transport equations, systems with actuator delay, Kuramoto–Sivashinsky-like and Korteweg–de Vries-like linear PDEs, and Navier–Stokes equations. We also cover the basics of motion planning and parameter-adaptive control for PDEs, as well as observer design with boundary sensing.

Short versions of a course on boundary control, based on preliminary versions of the book, were taught at the University of California, Santa Barbara (in Spring 2006) and at the University of California, Berkeley (in Fall 2007 and where the first author was invited to present the course as a Russell Severance Springer Professor of Mechanical Engineering).

Acknowledgments. This book is a tutorial summary of research that a number of gifted collaborators have contributed to: Andras Balogh, Weijiu Liu, Dejan Boskovic, Ole Morten Aamo, Rafael Vazquez, and Jennie Cochran. Aside from our collaborators, first and foremost we thank Petar Kokotovic, who read the manuscript from cover to cover and provided hundreds (if not thousands) of constructive comments and edits. Our very special thanks also go to Charles Kinney and Matt Graham. While developing the field of backstepping for PDEs, we have also enjoyed support from and interaction with Mihailo Jovanovic, Irena Lasiecka, Roberto Triggiani, Baozhu Guo, Alexandre Bayen, George Weiss, Michael Demetriou, Sanjoy Mitter, Jessy Grizzle, and Karl Astrom. For the support from the National Science Foundation that this research has critically depended on, we are grateful to Drs. Kishan Baheti, Masayoshi Tomizuka, and Mario Rotea. Finally, we wish to thank our loved ones, Olga, Katerina, Victoria, Alexandra, and Angela.

La Jolla, California
November, 2007

MIROSLAV KRSTIC
ANDREY SMYSHLYAEV

Chapter 1

Introduction

Fluid flows in aerodynamics and propulsion applications; plasmas in lasers, fusion reactors, and hypersonic vehicles; liquid metals in cooling systems for tokamaks and computers as well as in welding and metal casting processes; acoustic waves and water waves irrigation systems. Flexible structures in civil engineering applications, aircraft wings and helicopter rotors, astronomical telescopes, and in nanotechnology devices such as the atomic force microscope. Electromagnetic waves and quantum mechanical systems. Waves and "ripple" instabilities in thin film manufacturing and in flame dynamics. Chemical processes in process industries and in internal combustion engines.

For a control engineer, few areas of control theory are as physically motivated—but also as impenetrable—as control of partial differential equations (PDEs). Even "toy" problems such as heat equations and wave equations (neither of which are unstable) require the user to have a considerable background in PDEs and functional analysis before one can study the control design methods for these systems, particularly *boundary control* design. As a result, courses in control of PDEs are extremely rare in engineering programs. Control students are seldom trained in PDEs (let alone in control of PDEs) and are cut off from numerous physical applications in which they could be making contributions, either on the technological or on the fundamental level.

With this book we hope to help change this situation. We introduce methods which are easy to understand, require minimal background beyond calculus, and are easy to teach—even for instructors with only minimal exposure to the subject of control of PDEs.

1.1 Boundary Control

Control of PDEs comes in roughly two settings—depending on where the actuators and sensors are located—"in domain" control, where the actuation penetrates inside the domain of the PDE system or is evenly distributed everywhere in the domain (likewise with sensing) and "boundary" control, where the actuation and sensing are applied only through the boundary conditions. Boundary control is generally considered to be physically more realistic because actuation and sensing are nonintrusive (think, for example, of a fluid flow where actuation would normally be from the walls of the flow domain).[1] Boundary control is also generally considered to be the harder problem, because the "input operator" (the analog of the B matrix in the LTI (linear time invariant) finite-dimensional model $\dot{x} = Ax + Bu$) and the output operator (the analog of the C matrix in $y = Cx$) are unbounded operators. As a result of the greater mathematical difficulty, fewer methods have been developed over the years for boundary control problems of PDEs, and most books on control of PDEs either don't cover boundary control or dedicate only small fractions of their coverage to boundary control.

This book is devoted exclusively to boundary control. One reason is because this is the more realistic problem called for by many of the current applications. Another reason is because the method that this book pursues has so far been developed only for boundary control problems because of a natural fit between the method and the boundary control paradigm.

1.2 Backstepping

The method that this book develops for control of PDEs is the so-called *backstepping* control method. Backstepping is a particular approach to stabilization of dynamic systems and is particularly successful in the area of nonlinear control [97]. Backstepping is unlike any of the methods previously developed for control of PDEs. It differs from optimal control methods in that it sacrifices optimality (though it can achieve a form of "inverse optimality") for the sake of avoiding operator Riccati equations, which are very hard to solve for infinite- (or high-) dimensional systems such as PDEs. Backstepping is also different from pole placement methods because, even though its objective is stabilization, which is also the objective of the pole placement methods, backstepping does not pursue precise assignment of even a finite subset (let alone the entire infinite collection) of the PDE's eigenvalues. Instead, the backstepping method achieves Lyapunov stabilization, which is often achieved by collectively shifting all the eigenvalues in a favorable direction in the complex plane, rather than by assigning individual eigenvalues. As the reader will soon learn, this task can be achieved in a rather elegant way, where the control gains are easy to compute symbolically, numerically, and in some cases even explicitly.

1.3 A Short List of Existing Books on Control of PDEs

The area of control of infinite-dimensional systems—PDEs and delay systems—has been under development since at least the 1960s. Some of the initial efforts that laid the foundations of the field in the late 1960s and early 1970s were in optimal control and controllability of linear PDE systems. Controllability is a very challenging question for PDE systems because a rank test of the type $[b, Ab, A^2b, \ldots, A^{n-1}b]$ cannot be developed. It is also a fascinating topic because different classes of PDEs call for very different techniques. *Controllability* is an important topic because, even though it does not yield controllers robust to uncertainties in initial conditions and modeling errors, it answers the fundamental question of whether input functions can be found to steer a system from one state to another.

[1] "Body force" actuation of electromagnetic type is also possible, but it has low control authority and its spatial distribution typically has a pattern that favors the near-wall region.

Feedback design methods for PDEs, which do possess robustness to various uncertainties, include different varieties of *linear quadratic optimal control*, *pole placement*, and other ideas extended from finite-dimensional to infinite-dimensional systems.

Some of the books on control of distributed parameter systems that have risen to prominence as research references are those by Curtain and Zwart [44], Lasiecka and Triggiani [109], Bensoussan et al. [22], and Christofides [36]. Application-oriented books on control of PDEs have been dedicated to problems that arise from flexible structures [122, 103, 106, 17, 145] and from flow control [1, 64].

1.4 No Model Reduction in This Book

Model reduction often plays an important role in most methods for control design for PDEs. Different methods employ model reduction in different ways to extract a finite-dimensional subsystem to be controlled while showing robustness to neglecting the remaining infinite-dimensional dynamics in the design. The backstepping method does not employ model reduction—none is needed except at the implementation stage, where the integral operators for state feedback laws are approximated by sums and when PDE observers are discretized for numerical implementation.

1.5 Control Objectives for PDE Systems

One can pursue several different objectives in a control design for both ordinary differential equation (ODE) and PDE systems. If the system is already stable, a typical objective for feedback control would be to *improve performance*. Optimality methods are natural in such situations. Another control objective is *stabilization*. This book almost exclusively focuses on open-loop unstable PDE plants and delivers feedback laws of acceptable complexity which solve the stabilization problem without resorting to operator Riccati equations.

Stabilization of an equilibrium state is a special case of a control problem in which one wants to stabilize a desired motion, such as in *trajectory tracking*. Before pursuing the problem of designing a feedback law to stabilize a desired trajectory, one first needs to solve the problem of *trajectory generation* or *motion planning*, where an open-loop input signal needs to be found that generates the desired output trajectory in the case of a perfect initial condition (that is matched to the desired output trajectory). No book currently exists that treats motion planning or trajectory tracking for PDEs in any detail. These problems are at an initial stage of research. We dedicate Chapter 12 of this book to motion planning and trajectory tracking for PDEs. Our coverage is elementary and tutorial and it adheres to the same goal of developing easily computable and, whenever possible, explicit results for trajectory generation and tracking of PDEs.

1.6 Classes of PDEs and Benchmark PDEs Dealt with in This Book

In contrast to ODEs, no general methodology can be developed for PDEs, neither for analysis nor for control synthesis. This is a discouraging fact about this field but is also one of its

Table 1.1. *Categorization of PDEs in the book.*

	∂_t	∂_{tt}
∂_x	transport PDEs, delays	
∂_{xx}	parabolic PDEs, reaction-advection-diffusion systems	hyperbolic PDEs, wave equations
∂_{xxx}	Korteweg–de Vries	
∂_{xxxx}	Kuramoto–Sivashinsky and Navier–Stokes (Orr–Sommerfeld form)	Euler-Bernoulli and shear beams, Schrödinger, Ginzburg–Landau

most exciting features because it provides an endless stream of opportunities for learning and creativity. Two of the most basic categories of PDEs studied in textbooks are *parabolic* and *hyperbolic* PDEs, with standard examples being heat equations and wave equations. However, the reality is that there are many more classes and examples of PDEs that don't fit into these two basic categories. By browsing through our table of contents the reader will observe the large number of different classes of PDEs and the difficulty with fitting them into neat groups and subgroups. Table 1.1 shows one categorization of the PDEs considered in this book, organized by the highest derivative in time, ∂_t, and in space, ∂_x, that an equation contains. We are considering in this book equations with up to two derivatives in time, ∂_{tt}, and with up to four derivatives in space, ∂_{xxxx}. This is not an exhaustive list. One can encounter PDEs with even higher derivatives. For example, the Timoshenko beam model has four derivatives in both time and space (a backstepping design for this PDE is presented in [99, 100]).

All the PDEs listed in Table 1.1 are regarded as real-valued. However, some of these PDEs, particularly the Schrödinger equation and the Ginzburg–Landau equation, are commonly studied as complex-valued PDEs. In that case they are written in a form that includes only one derivative in time and two derivatives in space. Their coefficients are also complex-valued in that case. For this reason, even though they "look" like parabolic PDEs, they behave like oscillatory, hyperbolic PDEs. In fact, the (linear) Schrödinger equation is, in some ways, equivalent to the Euler–Bernoulli beam PDE.

The list in Table 1.1 is not exhaustive. Many other benchmark PDEs exist, particularly when one explores the possibility of these PDEs containing nonlinearities.

1.7 Choices of Boundary Controls

As indicated earlier, this book deals exclusively with boundary control. However, more then one option exists when it comes to boundary actuation. In some applications it is natural to actuate the boundary value of the state variable of the PDE (Dirichlet actuation). This is the case, for example, in flow control where microjects are used to actuate the boundary values of the velocity at the wall. In contrast, in other applications it is only natural to actuate the boundary value of the gradient of the state variable of the PDE (Neumann actuation). This is the case, for example, in thermal problems where one can actuate the heat flux (the derivative of the temperature) but not the temperature itself. All of our designs are implementable using both Dirichlet and Neumann actuation.

1.8 The Domain Dimension: 1D, 2D, and 3D Problems

PDE control problems are complex enough for domains of one dimension, such as, a string, a beam, a chemical tubular reactor, etc. They can be unstable, can have a large number of unstable eigenvalues (in the presence of external "feedback-type" forces that induce instability), and can be highly nontrivial to control. However, many physical PDE problems exist which evolve in two and three dimensions. It is true that some of them are dominated by phenomena evolving in one coordinate direction (while the phenomena in the other directions are stable and slow) but there exist some equations that are genuinely three-dimensional (3D). This is particularly the case with Navier–Stokes equations, where the full richness and realism of turbulent fluid behavior is exhibited only in three dimensions. We present one such problem: a boundary control design for a 3D Navier–Stokes flow in a channel.

2D and 3D problems are easier when the domain shape is "regular" in some way. For example, PDE control for a system whose domain is a rectangle or an annulus is much more readily tractable than for a problem in which the domain has an "amorphous" shape. In oddly shaped domains, particularly if the PDE is unstable, the control problem is formidable for any method. Unfortunately, the literature abounds with abstract control methods for 2D and 3D PDE systems on general domains, where the complexities are hidden behind neatly written Riccati equations, with no numerical results offered that show that such equations are efficiently solvable and that the closed-loop system works well and is robust to modeling uncertainties. One should realize that genuine 2D and 3D systems, particularly if unstable and on oddly shaped domains, are typically impossible to represent by reasonable low-order finite-dimensional approximations. Such systems, such as turbulent fluids in three dimensions around irregularly shaped bodies, truly require millions of differential equations to capture their dynamics faithfully and tens of thousands of equations to perform control design properly. Systems of this type are still beyond the reach of the control methodologies, numerical methods, and the computer hardware available today. The reader should not expect to find solutions to such problems (unstable PDEs on irregularly shaped 2D and 3D domains, with boundary control) in this book, and, for that matter, in any other book on control of PDEs. Currently such problems are still on the list of future topics in backstepping control research for PDEs.

A reasonable setup for boundary control, which is still "hard" but tractable, is a setup where one has an order of magnitude fewer control inputs than states. This is the case with *end point* boundary control in one dimension (covered throughout this book), boundary control of a part of a 1D boundary in a 2D problem (covered, for example, in [174]), and boundary actuation of a part of a 2D boundary in a 3D problem (covered in Chapter 11 on Navier–Stokes equations).

1.9 Observers

In addition to presenting the methods for boundary control design, we present the dual methods for observer design using boundary sensing. Virtually every one of our control designs for full state stabilization has an observer counterpart, although to avoid monotony we don't provide all those observers. The observer gains are easy to compute symbolically or even explicitly in some cases. They are designed in such a way that the observer error system is exponentially stabilized. As in the case of finite-dimensional observer-based control, a separation principle holds in the sense that a closed-loop system remains stable after a full state stabilizing feedback is replaced by a feedback that employs the observer state instead of the plant state.

1.10 Adaptive Control of PDEs

In addition to developing state estimators for PDEs, we also show approaches for developing parameter estimators—system *identifiers*—for PDEs. This is an extremely challenging problem, especially for unstable PDEs. We give examples of adaptive control systems in which unstable PDEs with unknown parameters are controlled using parameter estimators supplied by identifiers and using state estimators supplied by observers. We provide example proofs of stability—a striking result, given that the infinite-dimensional plant is unstable, only a scalar output is measured, and only a scalar input is actuated. The area of adaptive control of PDEs is in its infancy. We present only reasonably "simple" results of tutorial value. A whole array of more advanced results and different methods is contained in our papers. Adaptive control of PDEs is a wide open, fertile area for future research.

1.11 Nonlinear PDEs

Currently, virtually no methods exist for boundary control of nonlinear PDEs. Several results are available that apply to nonlinear PDEs which are neutrally stable, where the nonlinearity plays no destabilizing role, that is, where a simple boundary feedback law needs to be carefully designed so that the PDE is changed from neutrally stable to asymptotically stable. However, no advanced control designs exist for nonlinear PDEs which are open-loop unstable, where a sophisticated control Lyapunov function of non-quadratic type needs to be constructed to achieve closed-loop stability. Even though the focus of this book is on linear PDEs, we introduce initial ideas and current results that exist for stabilization of nonlinear PDEs.

1.12 Organization of the Book

We begin, in Chapter 2, with an introduction to basic Lyapunov stability ideas for PDEs. Backstepping is a method which employs a change of variable and boundary feedback to make an unstable system behave like another, stable system, where the destabilizing terms from the plant have been eliminated. With the backstepping method an entire class of PDEs can be transformed into a single representative and desirable PDE from that class. For example, all parabolic PDEs (including unstable reaction-diffusion equations) can be transformed into a heat equation. Therefore, to study closed-loop stability of systems controlled by backstepping, one should focus on the stability properties of certain basic (and stable) PDEs, such as the heat equation, a wave equation with appropriate damping, and other stable examples from other classes of PDEs. For this reason, our coverage of stability of PDEs in Chapter 2 focuses on such basic PDEs, highlighting the role of spatial norms (L_2, H_1, and so on), the role of the Poincaré, Agmon, and Sobolev inequalities, the role of integration by parts in Lyapunov calculations, and the distinction between energy boundedness and pointwise (in space) boundedness. The reader should immediately appreciate that it is possible to determine whether a PDE behaves desirably even if one has not yet learned how (or will never be able) to solve it.

In Chapter 3 we introduce the concepts of eigenvalues and eigenfunctions and the basics of finding the solutions of PDEs analytically.

In Chapter 4 we introduce the backstepping method. This is done for the class of parabolic PDEs. Our main "tutorial tool" is the reaction-diffusion PDE example

$$u_t(x,t) = u_{xx}(x,t) + \lambda u(x,t)\,,$$

which is evolving on the spatial interval $x \in (0, 1)$, with one uncontrolled boundary condition at $x = 0$,

$$u(0,t) = 0\,,$$

and with a control applied through Dirichlet boundary actuation at $x = 1$. The reaction term $\lambda u(x,t)$ in the PDE can cause open-loop instability. In particular, for a large positive λ this system can have arbitrarily many unstable eigenvalues. Using the backstepping method, we design the full-state feedback law,

$$u(1,t) = -\int_0^1 \lambda y \frac{I_1\left(\sqrt{\lambda(1-y^2)}\right)}{\sqrt{\lambda(1-y^2)}} u(y,t)dy\,,$$

where I_1 is a Bessel function, and show that this controller is exponentially stabilizing. This is a remarkable result, not only because such an explicit formula for a PDE control law is not achievable with any of the other previously developed methods for control of PDEs (note that the gain kernel formula is given explicitly both in terms of the spatial variable y and in terms of the parameter λ), but also because the simplicity of the formula and of the method itself makes this PDE control design easy to understand by users with very little PDE background, making it an ideal entry point for studying control of PDEs. The rest of Chapter 4 is dedicated to various generalizations and extensions within the class of parabolic PDEs.

In Chapter 5 we develop the observer design approach. The class of parabolic PDEs from Chapter 4 is used here to introduce the design ideas and tools of this approach, but such observers can be developed for all classes of PDEs in this book.

In Chapter 6 we consider Schrödinger and Ginzburg–Landau PDEs. They look like parabolic PDEs. For example, the linear Schrödinger equation looks exactly like a heat equation but its diffusion coefficient is imaginary and its state variable is complex valued. The designs for these equations follow easily after the designs from Chapter 4. An important example of an output feedback controller applied to a model of vortex shedding is presented.

Chapters 7, 8, and 9 deal with hyperbolic and "hyperbolic-like" equations—wave equations, beams, transport equations, and delay equations. For wave and beam equations, we introduce an important new "trick" that distinguishes the design for this class from the design for parabolic PDEs. This trick has to do with how to introduce damping into these systems where, as it turns out, damping (of "viscous" type) cannot be introduced directly using boundary control. In Chapter 9 we show a backstepping design for compensation of an actuator delay in a stabilizing control design for an unstable finite-dimensional plant. To achieve this, we treat the delay as a transport PDE.

In Chapter 10 we turn our attention to the "exotics"—PDEs with just one spatial derivative but with three, and even four, spatial derivatives—the Kuramoto–Sivashinsky and the Korteweg–de Vries equations (or their variants, to be precise).

Chapter 11 is the only chapter in this book that deals with a problem that is more than a benchmark problem. We present an application of backstepping to stabilization of a 3D high Reynolds number Navier–Stokes equation.

In Chapter 12 we introduce the basics of motion planning/trajectory generation for PDEs. For example, we consider how to move one end of a flexible beam to produce precisely the desired motion with the free end of the beam. An example is then given of how to combine the motion planning results from Chapter 12 with the feedback results from the other chapters to achieve trajectory tracking.

In Chapter 13 we introduce the key elements of adaptive control for parametrically uncertain PDEs.

Finally, in Chapter 14 we introduce the main idea for designing backstepping boundary controllers for nonlinear PDEs.

1.13 Why We Don't State Theorems

This book is intended for students being introduced to PDE control for the first time–not only the most mathematically inclined Ph.D. students, but also Masters students in control *engineering* who are aware of many applications for PDE control and enroll in a course with the primary intent of learning control algorithms (rather than primarily proofs), but who remain interested in the subject only if some of the technical details are "swept under the rug." Our presentation style is tuned to ease the latter students into the subjects of boundary control and backstepping design for PDEs, while also providing a more mathematically oriented student with plenty of food for thought in the form of challenging problems that other methods do not address effectively. Although theorems are not stated for the designs that we develop and whose stability we prove here, theorems *are* stated in our papers that have inspired the writing of this text. An astute reader will easily deduce the missing details from the choices of Lyapunov functions that we use and from boundary conditions that we impose, along with the help of the properties of our changes of variables. The spaces that our closed-loop systems are defined on, the assumptions needed on the initial conditions, and other technical details are all implicit in our stability presentation. The well posedness issues for our designs are nearly trivial because our "target systems" are basic PDEs which are not only well studied in the literature but also explicitly solvable. We give plenty of simulation results, which are probably more useful to most students than theorems. In our experience with teaching this course to a diverse engineering audience, we found that going through the material at a brisk pace, without stating theorems, but with highlighting the physical intuition, works better; in fact, this is the only approach that does work for our intended purpose of attracting interest in control of PDEs from a broader set of potential technology-oriented users.

1.14 Focus on Unstable PDEs and Feedback Design Difficulties

Despite the fact that a reader who uses mathematics for a living (particularly, who studies properties of solutions of PDEs such as existence, uniqueness, the function spaces to which the solutions belong, etc.) won't find our exposition fitting the formalisms preferred in his/her field of work, the book successfully focuses on problems that are challenging per today's standards. We deal primarily with open-loop unstable problems. Even within the class of hyperbolic PDEs, we identify classes of physically motivated *unstable* plants (Exercises 7.3 and 8.3) for which boundary control is feasible. An instructor using this text to ease his/her students into the field of control of PDEs can rest assured that the challenge of the control problems treated in the text is not eliminated by the accessible presentation style. Likewise, a mathematically inclined researcher will find plenty of problems to inspire future research in the directions that present greater challenges, particularly in the construction of feedback laws for stabilization of PDEs.

1.15 The Main Idea of Backstepping Control

A reader who wants to understand the main idea of backstepping design can go straight to Section 4.1, whereas the reader who is primarily intrigued by the name "backstepping" will find in Section 4.11 a (long) discussion of the origins of this term. The key ideas of backstepping design for PDEs are intricately linked to the key ideas behind "feedback linearization" techniques for control of nonlinear finite-dimensional systems.[2] Feedback linearization focuses on nonlinearities in the plant, including those whose effect on stability is potentially harmful. Feedback linearization entails two steps: the first is the construction of an invertible change of variables such that the system appears linear in the new variables (except for a nonlinearity which is "in the span" of the control input vector), and the second is the cancellation of this nonlinearity[3] and the assignment of desirable linear exponentially stable dynamics on the closed-loop system. Hence, the system is made to behave as an easy-to-analyze linear system in the new variables, while its behavior in the original variables is also predictably stable due to the fact that the change of variables has a well-defined inverse.

Our extension of this design framework to PDEs goes as follows. First, one identifies the undesirable term or terms in a PDE model. This task may be trivial when the open-loop

[2]For finite-dimensional nonlinear systems, "backstepping" is an extension of "feedback linearization," which provides design tools that endow the controller with robustness to uncertain parameters and functional uncertainties in the plant nonlinearities, robustness to external disturbances, and robustness to other forms of modeling errors, including some dynamic uncertainties.

[3]In contrast to standard feedback linearization, backstepping allows the flexibility to not necessarily cancel the nonlinearity. A nonlinearity may be kept if it is useful, or it may be dominated (rather than cancelled non-robustly) if it is potentially harmful and uncertain. A rather subtle use of the capability of the backstepping method to avoid the cancellation of useful terms is included in Section 6.2 the design of a controller for the suppression of vortex shedding in a Ginzburg–Landau PDE model.

plant is such that stability can be analyzed easily, for example, by computing the open-loop eigenvalues. Then one decides on a "target system" in which the undesirable terms are eliminated after the application of a change of variable and feedback, as in the case of feedback linearization. The change of variable is a key ingredient of the method. It shifts the system state by a Volterra operator (in the spatial variable of the PDE) acting on the same state. A Volterra operator is an integral operator, where the integral runs from 0 up to x as the upper limit (rather than being an integral over the entire spatial domain). The Volterra form of the shift of the state variable means that the state transformation is "triangular," as it is in the backstepping design for finite-dimensional nonlinear systems. This triangularity ensures the invertibility of the change of variable. One can also think of this change of variable as being "spatially causal." Another key ingredient of the backstepping method is the boundary feedback obtained from the Volterra transformation. The transformation alone is obviously not capable of eliminating the undesirable terms. This transformation only "brings them to the boundary," where the boundary feedback controller can cancel them (or dominate them).

The gain function of the boundary controller is defined by the kernel of the Volterra transformation. As the reader will find in Chapter 4, this kernel satisfies an interesting PDE which is linear and easily solvable. The simplicity of finding the control gains is one of the main benefits of the backstepping method. The standard methods for control of PDEs, which are "PDE extensions" of linear-quadratic-optimal methods for finite-dimensional systems obtain their gains through the solution of Riccati equations—quadratic operator-valued equations which are in general hard to solve.

We emphasize, however, that the numerical advantage of backstepping is just part of its appeal. Another advantage is the conceptual elegance of using a feedback transformation that eliminates exactly the undesirable part of the PDE (or adds a part that is missing, such as damping), while leaving the PDE in a form which is familiar to the user, where physical intuition can be used in shaping the closed-loop dynamics.

Backstepping represents a major departure from the "one-size-fits-all" and "extending the results from finite-dimensional to infinite-dimensional systems" philosophies that drive some of the standard approaches in the field of PDE control. In the spirit of PDEs, where each class calls for a special treatment, backstepping requires the designer to physically understand what the desired, "target" behavior within the given class is and to develop a transformation capable of achieving such behavior. This requires a designer who is capable of more than just "turning the crank on a method," though this capability is more in the realm of creatively combining some basic ideas from physics and control design rather than the realm of very advanced mathematics. For the extra effort, the designer is rewarded with controllers and closed-loop systems whose properties are easy to understand and whose gains are easy to compute and tune.

Backstepping contradicts the conventional view that PDEs are "much harder to control than ODEs." If the only important feature of PDEs was their high dimension, then this statement would be true. However, there is a structure to PDEs. The role which the spatial derivatives play in this structure is crucial. If one approaches a particular PDE with a custom design, one can arrive at control designs of great simplicity—including even explicit formulae for control gains, as the reader shall see in Chapter 4.

1.16 Emphasis on Problems in One Dimension

The reader will observe a relative absence of problems in dimensions higher than one. Except for Chapter 11, where we treat linearized Navier–Stokes equations in three dimensions, the book focuses on PDEs in one dimension. Even in Chapter 11, the 3D problem dealt with is in a very special geometry, where the flow is between two parallel plates. Instinctively, one might view the absence of results for general geometries in two and three dimensions as a shortcoming. However, one should understand that in general geometries for PDEs one can only state abstract results. If we were to deal with general geometries, the clarity, elegance, and physical insight that we strive to achieve in this text would have to be abandoned. Noncomputable feedback laws (and even "potentially computable" feedback laws, whose implementation would require further research to deduce the implementation details from "existence" results) are not what this book is about. Hence, we concentrate mainly on 1D problems, where the challenges are primarily drawn from the instability of the plant and from the noncollocation of the control input and the source of instability.

1.17 Unique to This Book: Elements of Adaptive and Nonlinear Designs for PDEs

While our coverage of adaptive and nonlinear control designs is limited to examples (to maintain the tutorial and accessible character of the book), adaptive and nonlinear control are important elements of this book, as many of our designs for linear PDEs are "alternatives" to existing designs (alternatives that perhaps offer significant advantages), whereas the adaptive and nonlinear designs are the first and only methods available for some of the PDE problems considered. Our view is that the state of the art in adaptive control for PDEs before the introduction of backstepping was comparable to the state of the art of adaptive control for ODEs in the late 1960s, when the focus was on relative degree one (and related) problems. Similarly, the current state of the art in control of nonlinear PDEs is comparable to where the state of nonlinear control for ODEs was in the late 1970s. The backstepping idea that we introduce for designing nonlinear controllers and nonquadratic Lyapunov functions is intended to advance the nonlinear control for PDEs to where the state of nonlinear control for ODEs was in, roughly, the early 1990s. The idea builds upon the Volterra transformations employed in the linear parts of the book, but is generalized to nonlinear Volterra series. Our use of Volterra series is different than their conventional usage in representation theory for nonlinear ODEs, where Volterra series in time are employed. We use Volterra series in space to represent plant (static) nonlinearities and nonlinear (static, full-state) feedback laws rather than input-output dynamics.

1.18 How to Teach from This Book

Most of the material from this book was covered in a one-quarter course, MAE 287: Distributed Parameter Systems, at the University of California, San Diego in Fall 2005.

An instructor can cover about one chapter per week and be done with most of the book in one quarter (skipping Chapters 10, 13, and 14, for example); the entire book can be covered in one semester. The book contains many examples and simulation results to motivate the students through the material. Homework exercises are included. To obtain a solutions manual, the instructor should contact the authors at krstic@ucsd.edu.

Chapter 2

Lyapunov Stability

Before we venture forth into the study of stability analysis tools for PDEs, let us recall some of the basics of stability analysis for ODEs. Since we study only *linear* PDEs in this book, only the linear ODE stability theory is of relevance here.

An ODE

$$\dot{z} = Az \tag{2.1}$$

with $z \in \mathbb{R}^n$ is said to be exponentially stable at the equilibrium $z = 0$ if there exist positive constants M and α such that

$$\|z(t)\| \leq M\mathrm{e}^{-\alpha t}\|z(0)\| \quad \text{for all } t \geq 0\,, \tag{2.2}$$

where $\|\cdot\|$ denotes one of the equivalent vector norms, for example, the 2-norm.

This is a *definition* of stability and is not practical as a *test* of stability. One (necessary and sufficient) test that guarantees exponential stability is to verify that all the eigenvalues of the matrix A have negative real parts. However, this test is not always practical.

An alternative test, which can be used for studying the system robustness and which also extends to nonlinear ODEs, is the Lyapunov stability test. The system (2.1) is exponentially stable in the sense of definition (2.2) if and only if for any positive definite $n \times n$ matrix Q there exists a positive definite and symmetrical matrix P such that

$$PA + A^T P = -Q\,. \tag{2.3}$$

Along with this test comes the concept of a Lyapunov function $V = x^T Px$, which is positive definite and whose derivative $\dot{V} = -x^T Qx$ is negative definite.

The point of the Lyapunov method is finding the solution P to the Lyapunov matrix equation (2.3). This analysis paradigm extends to infinite-dimensional systems such as PDEs, but only on the abstract level. Even without the difficulties associated with solving an (infinite-dimensional) operator equation such as (2.3), one has to carefully consider the question of the definition of stability (2.2).

In a finite dimension, the vector norms are "equivalent." No matter which norm one uses for $\|\cdot\|$ in (2.2) (for example, the 2-norm, 1-norm, or ∞-norm), one can get exponential stability in the sense of any other vector norm. What changes are the constants M and α in the inequality (2.2).

For PDEs the situation is quite different. Since the state space is infinite-dimensional, for example, the state index i in $z_1, z_2, \ldots, z_i, \ldots, z_n$ is replaced by a continuous spatial variable x (in a PDE that evolves on a 1D domain), the state space is not a Euclidean space but a function space, and likewise, the state norm is not a vector norm but a function norm. Unfortunately, norms on function spaces are not equivalent; i.e., bounds on the state in terms of the L_1-, L_2-, or L_∞-norm in x do not follow from one another. To make matters worse, other natural choices of state norms for PDEs exist which are not equivalent with L_p-norms. Those are the so-called *Sobolev* norms, examples of which are the H_1- and H_2-norms (not to be confused with Hardy space norms in robust control for ODE systems), and which, roughly, are the L_2-norms of the first and second derivative, respectively, of the PDE state.

With such a variety of choices, such lack of generality in the meaning of the stability results, and such idiosyncrasy of the PDE classes being studied, general Lyapunov stability tests for PDEs offer almost no practical value. Instead, one has to develop an understanding of the relationships between functional norms and gain experience with deriving "energy estimates" in different norms.

In this chapter we try to give the reader a flavor of some of the main issues that arise in deriving exponential stability estimates for parabolic PDEs in one dimension. In the rest of the book we also consider hyperbolic PDEs and higher-dimensional domains.

2.1 A Basic PDE Model

Before introducing stability concepts, we develop a basic "nondimensionalized" PDE model, which will be the starting point for many of the analysis and control design considerations in this book.

Consider a thermally conducting (metal) rod (Figure 2.1) of length L whose temperature $T(\xi, \tau)$ is a function of the spatial variable ξ and time τ. The initial temperature distribution is $T(\xi)$ and the ends of the rod are kept at constant temperatures T_1 and T_2. The evolution of the temperature profile is described by the heat equation[4]

$$T_\tau(\xi, \tau) = \varepsilon T_{\xi\xi}(\xi, \tau)\,, \tag{2.4}$$

$$T(0, \tau) = T_1\,, \tag{2.5}$$

$$T(1, \tau) = T_2\,, \tag{2.6}$$

$$T(\xi, 0) = T_0(\xi)\,. \tag{2.7}$$

Here ε denotes the thermal diffusivity and T_τ and $T_{\xi\xi}$ are the partial derivatives with respect to time and space.

Our objective is to write this equation in nondimensional variables that describe the error between the unsteady temperature and the equilibrium profile of the temperature. This is done as follows:

1. We scale ξ to normalize the length:

$$x = \frac{\xi}{L}\,, \tag{2.8}$$

[4]While in physical heat conduction problems it is more appropriate to assume that the heat flux T_ξ (rather than the temperature T itself) is held constant at the boundaries, for simplicity we proceed with the boundary conditions as in (2.5) and (2.6).

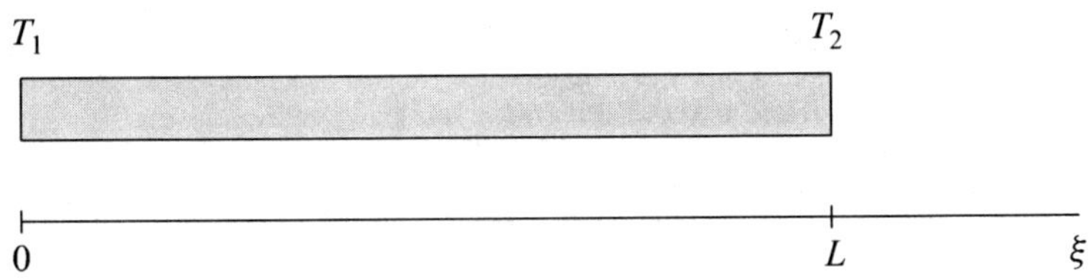

Figure 2.1. *A thermally conducting rod.*

which gives

$$T_\tau(x,\tau) = \frac{\varepsilon}{L^2} T_{xx}(x,\tau)\,, \tag{2.9}$$
$$T(0,\tau) = T_1\,, \tag{2.10}$$
$$T(1,\tau) = T_2\,. \tag{2.11}$$

2. We scale time to normalize the thermal diffusivity:

$$t = \frac{\varepsilon}{L^2}\tau\,, \tag{2.12}$$

which gives

$$T_t(x,t) = T_{xx}(x,t)\,, \tag{2.13}$$
$$T(0,t) = T_1\,, \tag{2.14}$$
$$T(1,t) = T_2\,. \tag{2.15}$$

3. We introduce the new variable

$$w = T - \bar{T}\,, \tag{2.16}$$

where $\bar{T}(x) = T_1 + x(T_2 - T_1)$ is the steady-state profile and is a solution to the two-point boundary-value ODE

$$\bar{T}''(x) = 0\,, \tag{2.17}$$
$$\bar{T}(0) = T_1\,, \tag{2.18}$$
$$\bar{T}(1) = T_2\,. \tag{2.19}$$

Finally, we obtain

$$w_t = w_{xx}\,, \tag{2.20}$$
$$w(0) = 0\,, \tag{2.21}$$
$$w(1) = 0\,, \tag{2.22}$$

where the initial distribution of the temperature fluctuation is $w_0 = w(x,0)$. Note that here, and throughout the rest of the book, for clarity of presentation we drop the dependence on time and spatial variable where it does not lead to confusion; i.e., by w, $w(0)$ we always mean $w(x,t)$, $w(0,t)$, respectively, unless specifically stated otherwise.

The following are the basic types of boundary conditions for PDEs in one dimension:

- Dirichlet: $w(0) = 0$ (fixed temperature at $x = 0$).
- Neumann: $w_x(0) = 0$ (fixed heat flux at $x = 0$).
- Robin: $w_x(0) + qw(0) = 0$ (mixed).

Throughout this text we will be studying problems with all three types of boundary conditions.

2.2 Lyapunov Analysis for a Heat Equation in Terms of "L_2 Energy"

Consider the heat equation

$$w_t = w_{xx}\,, \tag{2.23}$$
$$w(0) = 0\,, \tag{2.24}$$
$$w(1) = 0\,. \tag{2.25}$$

The basic question we want to answer is whether this system is exponentially stable in the sense of an L_2-norm of the state $w(x,t)$ with respect to the spatial variable x. It is true that this particular PDE can be solved in closed form (see the next chapter), and the stability properties can be analyzed directly from the explicit solution. Moreover, from the physical point of view, this system clearly cannot be unstable since there is no heat generation (reaction terms) in the equation. However, for more complex problems the explicit solution usually cannot be found, and even physical considerations may not be conclusive. Thus, it is important to develop a tool for analyzing the stability of such PDEs without actually solving them.

Consider the Lyapunov function[5]

$$V(t) = \frac{1}{2}\int_0^1 w^2(x,t)\,dx. \tag{2.26}$$

Let us calculate the time derivative of V:

$$\begin{aligned}
\dot{V} = \frac{dV}{dt} &= \int_0^1 w(x,t)w_t(x,t)dx \quad \text{(applying the chain rule)} \\
&= \int_0^1 ww_{xx}dx \qquad \text{(from (2.23))} \\
&= ww_x|_0^1 - \int_0^1 w_x^2 dx \quad \text{(integration by parts)} \\
&= -\int_0^1 w_x^2 dx\,.
\end{aligned} \tag{2.27}$$

[5] Strictly speaking, this is a functional, but we refer to it simply as a "Lyapunov function" throughout the book.

The time derivative of V shows that it is bounded. However, it is not clear whether or not V goes to zero because (2.27) depends on w_x and not on w, so one cannot express the right-hand side of (2.27) in terms of V.

Let us first recall two very useful inequalities:

Young's Inequality (special case):

$$ab \le \frac{\gamma}{2}a^2 + \frac{1}{2\gamma}b^2 . \tag{2.28}$$

Cauchy–Schwarz Inequality:

$$\int_0^1 uw\,dx \le \left(\int_0^1 u^2\,dx\right)^{1/2} \left(\int_0^1 w^2\,dx\right)^{1/2} . \tag{2.29}$$

The following lemma establishes the relationship between the L_2-norms of w and w_x.

Lemma 2.1 (Poincaré Inequality). *For any w, continuously differentiable on* $[0, 1]$,

$$\begin{aligned} \int_0^1 w^2\,dx &\le 2w^2(1) + 4\int_0^1 w_x^2\,dx , \\ \int_0^1 w^2\,dx &\le 2w^2(0) + 4\int_0^1 w_x^2\,dx . \end{aligned} \tag{2.30}$$

Remark 2.2. The inequalities (2.30) are conservative. The tight version of (2.30) is

$$\int_0^1 (w(x) - w(0))^2\,dx \le \frac{4}{\pi^2}\int_0^1 w_x^2\,dx , \tag{2.31}$$

which is sometimes called "a variation of Wirtinger's inequality." The proof of (2.31) is far more complicated than the proof of (2.30) and is given in [69].

Proof of Lemma 2.1.

$$\begin{aligned} \int_0^1 w^2\,dx &= xw^2|_0^1 - 2\int_0^1 xww_x\,dx \quad \text{(integration by parts)} \\ &= w^2(1) - 2\int_0^1 xww_x\,dx \\ &\le w^2(1) + \frac{1}{2}\int_0^1 w^2dx + 2\int_0^1 x^2w_x^2\,dx . \end{aligned}$$

Subtracting the second term from both sides of the inequality, we get the first inequality in (2.30):

$$\begin{aligned}\frac{1}{2}\int_0^1 w^2 dx &\le w^2(1) + 2\int_0^1 x^2 w_x^2 dx \\ &\le w^2(1) + 2\int_0^1 w_x^2 dx\,. \end{aligned} \tag{2.32}$$

The second inequality in (2.30) is obtained in a similar fashion. □

We now return to equation (2.27). Using the Poincaré inequality along with boundary conditions, we get

$$\dot{V} = -\int_0^1 w_x^2 dx \le -\frac{1}{4}\int_0^1 w^2 \le -\frac{1}{2}V \tag{2.33}$$

which, by the basic comparison principle for first-order differential inequalities, implies that the energy decay rate is bounded by

$$V(t) \le V(0)e^{-t/2} \tag{2.34}$$

or by

$$\|w(t)\| \le e^{-t/4}\|w_0\|\,, \tag{2.35}$$

where

$$w_0(x) = w(x,0)$$

is the initial condition and $\|\cdot\|$ denotes the L_2-norm of a function of x, namely,

$$\|w(t)\| = \left(\int_0^1 w(x,t)^2 dx\right)^{1/2}. \tag{2.36}$$

Thus, the system (2.23)–(2.25) is exponentially stable in L_2.

The dynamic response of the PDE in Figure 2.2 demonstrates the stability result that we have just established. Even though the solution starts from a nonsmooth initial condition, it rapidly smoothes out. This "instant smoothing" effect is the characteristic feature of the heat equation. Our major message here is that by using Lyapunov tools we predicted the overall decay of the solution without knowledge of the exact solution $w(x,t)$ for a specific initial condition $w_0(x)$.

For PDEs, the L_2 form of stability in (2.35) is just one of the many possible (nonequivalent) forms of stability, and the Lyapunov function (2.26) is just one of the many possible choices, a well known feature of the Lyapunov method for ODEs. Nevertheless, the L_2 stability, quantified by (2.26) and (2.35), is usually the easiest one to prove for a vast majority of PDEs, and an estimate of the form (2.35) is often needed before proceeding to study stability in higher norms.

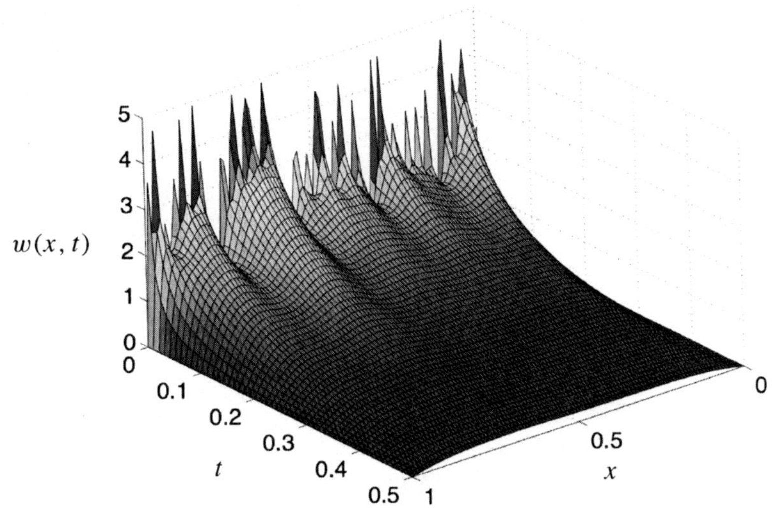

Figure 2.2. *Response of a heat equation to a nonsmooth initial condition.*

2.3 Pointwise-in-Space Boundedness and Stability in Higher Norms

We established that

$$\|w\| \to 0 \text{ as } t \to \infty,$$

but this does not imply that $w(x,t)$ goes to zero for all x. There could be "unbounded spikes" for some x along the spatial domain (on a set of measure zero) that do not contribute to the L_2-norm; however, Figure 2.2 shows that this is unlikely to occur, as our analysis to follow will demonstrate.

It would be thus desirable to show that

$$\max_{x\in[0,1]} |w(x,t)| \le e^{-\frac{t}{4}} \max_{x\in[0,1]} |w(x,0)|, \tag{2.37}$$

namely, stability in the spatial L_∞-norm. This is possible only in some special cases, using problem specific considerations, and therefore is not worth our attention in a text that focuses on basic but generally applicable tools. However, it is easy to show a more restrictive result than (2.37), given by

$$\max_{x\in[0,1]} |w(x,t)| \le K e^{-\frac{t}{2}} \|w(x,0)\|_{H_1}, \tag{2.38}$$

for some $K > 0$, where the H_1-norm is defined by

$$\|w\|_{H_1} := \int_0^1 w^2 dx + \int_0^1 w_x^2 dx. \tag{2.39}$$

Remark 2.3. The H_1-norm can be defined in different ways, but the definition given above suits our needs. Note also that by using the Poincaré inequality, it is possible to drop the first integral in (2.39) for most problems.

Before we proceed to prove (2.38), we need the following result.

Lemma 2.4 (Agmon's Inequality). *For a function $w \in H_1$, the following inequalities hold:*

$$\boxed{\begin{aligned} \max_{x\in[0,1]} |w(x,t)|^2 &\le w(0)^2 + 2\|w(t)\|\,\|w_x(t)\|\,, \\ \max_{x\in[0,1]} |w(x,t)|^2 &\le w(1)^2 + 2\|w(t)\|\,\|w_x(t)\|\,. \end{aligned}} \tag{2.40}$$

Proof.

$$\begin{aligned} \int_0^x w w_x dx &= \int_0^x \partial_x \frac{1}{2} w^2 dx \\ &= \frac{1}{2} w^2|_0^x \\ &= \frac{1}{2} w(x)^2 - \frac{1}{2} w(0)^2. \end{aligned} \tag{2.41}$$

Taking the absolute value on both sides and using the triangle inequality gives

$$\frac{1}{2}|w(x)^2| \le \int_0^x |w||w_x|dx + \frac{1}{2} w(0)^2. \tag{2.42}$$

Using the fact that an integral of a positive function is an increasing function of its upper limit, we can rewrite the last inequality as

$$|w(x)|^2 \le w(0)^2 + 2\int_0^1 |w(x)||w_x(x)|dx. \tag{2.43}$$

The right-hand side of this inequality does not depend on x, and therefore

$$\max_{x\in[0,1]} |w(x)|^2 \le w(0)^2 + 2\int_0^1 |w(x)||w_x(x)|dx. \tag{2.44}$$

Using the Cauchy–Schwarz inequality we get the first inequality of (2.40). The second inequality is obtained in a similar fashion. □

The simplest way to prove (2.38) is to use the following Lyapunov function:

$$V_1 = \frac{1}{2}\int_0^1 w^2\,dx + \frac{1}{2}\int_0^1 w_x^2\,dx\,. \tag{2.45}$$

The time derivative of (2.45) is given by

$$\begin{aligned} \dot{V}_1 &\le -\|w_x\|^2 - \|w_{xx}\|^2 \le -\|w_x\|^2 \\ &\le -\frac{1}{2}\|w_x\|^2 - \frac{1}{2}\|w_x\|^2 \\ &\le -\frac{1}{8}\|w\|^2 - \frac{1}{2}\|w_x\|^2 \quad \text{(using (2.33))} \\ &\le -\frac{1}{4}V_1. \end{aligned}$$

Therefore,

$$\|w\|^2 + \|w_x\|^2 \le e^{-t/2}\left(\|w_0\|^2 + \|w_{0,x}\|^2\right), \tag{2.46}$$

and using Young's and Agmon's inequalities, we get

$$\begin{aligned}\max_{x\in[0,1]} |w(x,t)|^2 &\le 2\|w\|\|w_x\| \\ &\le \|w\|^2 + \|w_x\|^2 \\ &\le e^{-t/2}\left(\|w_0\|^2 + \|w_{x,0}\|^2\right). \end{aligned} \tag{2.47}$$

We have thus showed that

$$w(x,t) \to 0 \text{ as } t \to \infty$$

for all $x \in [0, 1]$.

In the example that we consider next, we advance from the simple heat equation studied so far to considering the stability problem for an advection-diffusion PDE.

Example 2.5 Consider the *diffusion-advection* equation

$$w_t = w_{xx} + w_x\,, \tag{2.48}$$

$$w_x(0) = 0\,, \tag{2.49}$$

$$w(1) = 0\,. \tag{2.50}$$

Let us show the L_2 stability of this system by employing the tools introduced above.

Using the Lyapunov function (2.26), we get

$$\begin{aligned}\dot V &= \int_0^1 w w_t dx \\ &= \int_0^1 w w_{xx} dx + \int_0^1 w w_x dx \\ &= w w_x|_0^1 - \int_0^1 w_x^2 dx + \int_0^1 w w_x dx \qquad \text{(integration by parts)} \\ &= -\int_0^1 w_x^2 dx + \frac{1}{2} w^2|_0^1 \\ &= -\int_0^1 w_x^2 dx - \frac{1}{2} w^2(0)\,.\end{aligned}$$

Finally, using the Poincaré inequality (2.30), we get

$$\dot V \le -\frac{1}{4}\|w\|^2 \le -\frac{1}{2} V\,, \tag{2.51}$$

proving the exponential stability in L_2 norm,

$$\|w(t)\| \le \mathrm{e}^{-t/4}\|w_0\|. \qquad \diamond$$

2.4 Notes and References

An excellent coverage of the Lyapunov stability method for both linear and nonlinear finite-dimensional systems is given in [82]. Efforts towards developing Lyapunov theory for infinite-dimensional systems are made in [71, 182].

It might seem from this chapter that our entire concept of Lyapunov theory for PDEs reduces to manipulating basic norms and estimating their evolution in time. It might appear that we are not constructing any nontrivial Lyapunov functions but using only the "diagonal" Lyapunov functions that do not involve any "cross-terms." This is actually not the case with the remainder of the book. While in this chapter we have studied only Lyapunov functions that are plain spatial norms of functions, in the remaining chapters we are going to construct changes of variables for the PDE states, which are going to facilitate our analysis in a way so that we can use Lyapunov results for system norms. The Lyapunov functions will employ the norms of the *transformed* state variables, which means that in the original PDE state our Lyapunov functions will be complex, sophisticated constructions that include nondiagonal and cross-term effects.

Exercises

2.1. Prove the second inequalities in (2.30) and (2.40).

2.2. Consider the heat equation

$$w_t = w_{xx}$$

for $x \in (0, 1)$ with the initial condition $w_0(x) = w(x, 0)$ and boundary conditions

$$w_x(0) = 0$$
$$w_x(1) = -\frac{1}{2} w(1) .$$

Show that

$$\|w(t)\| \leq e^{-\frac{t}{4}} \|w_0\| .$$

2.3. Consider the Burgers equation

$$w_t = w_{xx} - w w_x$$

for $x \in (0, 1)$ with the initial condition $w_0(x) = w(x, 0)$ and boundary conditions

$$w(0) = 0 ,$$
$$w_x(1) = -\frac{1}{6} \left(w(1) + w^3(1) \right) .$$

Show that

$$\|w(t)\| \leq e^{-\frac{t}{4}} \|w_0\| .$$

Hint: Complete the squares.

Chapter 3

Exact Solutions to PDEs

In general, seeking explicit solutions to PDEs is a hopeless pursuit. One class of PDEs for which the closed-form solutions can be found is linear PDEs with constant coefficients. The solution not only provides us with an exact formula for a given initial condition, but also gives us insight into the spatial structure and the temporal evolution of the PDE.

3.1 Separation of Variables

Consider the diffusion equation which includes a reaction term

$$u_t = u_{xx} + \lambda u \tag{3.1}$$

with boundary conditions

$$u(0) = 0\,, \tag{3.2}$$

$$u(1) = 0\,, \tag{3.3}$$

and initial condition $u(x, 0) = u_0(x)$. Let us find the solution to this system and determine for which values of the parameter λ this system is unstable.

The most frequently used method for obtaining solutions to PDEs with constant coefficients is the method of separation of variables (the other common method employs the Laplace transform). Let us assume that the solution $u(x, t)$ can be written as a product of a function of space and a function of time,

$$u(x, t) = X(x)T(t)\,. \tag{3.4}$$

If we substitute the solution (3.4) into the PDE (3.1), we get

$$X(x)\dot{T}(t) = X''(x)T(t) + \lambda X(x)T(t). \tag{3.5}$$

Gathering the like terms on the opposite sides yields

$$\frac{\dot{T}(t)}{T(t)} = \frac{X''(x) + \lambda X(x)}{X(x)}. \tag{3.6}$$

Since the function on the left depends only on time, and the function on the right depends only on the spatial variable, the equality can hold only if both functions are constant. Let us denote this constant by σ. We then get two ODEs:

$$\dot{T} = \sigma T \tag{3.7}$$

with initial condition $T(0) = T_0$, and

$$X'' + (\lambda - \sigma)X = 0 \tag{3.8}$$

with boundary conditions $X(0) = X(1) = 0$ (they follow from the PDE boundary conditions). The solution to (3.7) is given by

$$T = T_0 e^{\sigma t}. \tag{3.9}$$

The solution to (3.8) has the form

$$X(x) = A\sin(\sqrt{\lambda - \sigma}x) + B\cos(\sqrt{\lambda - \sigma}x), \tag{3.10}$$

where A and B are constants that should be determined from the boundary conditions. We have

$$X(0) = 0 \Rightarrow B = 0\,,$$
$$X(1) = 0 \Rightarrow A\sin(\sqrt{\lambda - \sigma}) = 0\,.$$

The last equality can hold only if $\sqrt{\lambda - \sigma} = \pi n$ for $n = 0, 1, 2, \ldots$, so that

$$\sigma = \lambda - \pi^2 n^2, \qquad n = 0, 1, 2, \ldots. \tag{3.11}$$

Substituting (3.9) and (3.10) into (3.4) yields

$$u_n(x,t) = T_0 A_n e^{(\lambda - \pi^2 n^2)t}\sin(\pi n x), \qquad n = 0, 1, 2, \ldots. \tag{3.12}$$

For linear PDEs, the sum of particular solutions is also a solution (the principle of superposition). Therefore the formal general solution of (3.1)–(3.3) is given by

$$u(x,t) = \sum_{n=1}^{\infty} C_n e^{(\lambda - \pi^2 n^2)t}\sin(\pi n x)\,, \tag{3.13}$$

where $C_n = A_n T_0$.

The solution in the form (3.13) is sufficient for complete stability analysis of the PDE. If we are interested in the exact response to a particular initial condition, then we should determine the constants C_n. To do this, let us set $t = 0$ in (3.13) and multiply both sides of the resulting equality by $\sin(\pi m x)$:

$$u_0(x)\sin(\pi m x) = \sin(\pi m x)\sum_{n=1}^{\infty} C_n \sin(\pi n x)\,. \tag{3.14}$$

Then, using the identity

$$\int_0^1 \sin(\pi m x)\sin(\pi n x)dx = \left\{ \begin{array}{c} 1/2 \;\; n = m \\ 0 \;\; n \neq m \end{array} \right\}, \tag{3.15}$$

we get

$$C_m = \frac{1}{2}\int_0^1 u_0(x)\sin(\pi n x)dx\,. \tag{3.16}$$

Substituting this expression into (3.13), we get

$$\boxed{u(x,t) = 2\sum_{n=1}^{\infty} e^{(\lambda-\pi^2 n^2)t}\sin(\pi n x)\int_0^1 \sin(\pi n x)u_0(x)dx\,.} \tag{3.17}$$

Even though we obtained this solution formally, it can be proved that this is indeed a well-defined solution in the sense that it is unique, has continuous spatial derivatives up to second order, and depends continuously on the initial data.[6] Let us look at the structure of this solution. It consists of the following elements:

- eigenvalues: $\lambda - \pi^2 n^2$.
- eigenfunctions: $\sin(\pi n x)$.
- effect of initial conditions: $\int_0^1 \sin(\pi n x)u_0(x)dx$.

The largest eigenvalue, $\lambda - \pi^2$ ($n = 1$), indicates the rate of growth or decay of the solution. We can see that the plant is stable for $\lambda \leq \pi^2$ and unstable otherwise. After the transient response due to the initial conditions, the profile of the state will be proportional to the first eigenfunction $\sin(\pi x)$, since other modes decay much faster.

Sometimes it is possible to use the method of separation of variables to determine the stability properties of the plant even though the complete closed-form solution cannot be obtained, as in the following example.

Example 3.1 Let us find the values of the parameter g, for which the system

$$u_t = u_{xx} + g u(0)\,, \tag{3.18}$$

$$u_x(0) = 0\,, \tag{3.19}$$

$$u(1) = 0 \tag{3.20}$$

is unstable. This example is motivated by the model of thermal instability in solid propellant rockets, where the term $gu(0)$ is roughly the burning of the propellant at one end of the fuel chamber.

Using the method of separation of variables, we set $u(x,t) = X(x)T(t)$, and (3.18) gives

$$\frac{\dot{T}(t)}{T(t)} = \frac{X''(x) + gX(0)}{X(x)} = \sigma\,. \tag{3.21}$$

[6] The proof is standard and can be found in many PDE texts; e.g., see [40].

Hence, $T(t) = T(0)e^{\sigma t}$, whereas the solution of the ODE for X is given by

$$X(x) = A\sinh(\sqrt{\sigma}x) + B\cosh(\sqrt{\sigma}x) + \frac{g}{\sigma}X(0). \tag{3.22}$$

Here the last term is a particular solution of a nonhomogeneous ODE (3.21). Now we find the constant B in terms of $X(0)$ by setting $x = 0$ in the above equation. This gives $B = X(0)(1 - g/\sigma)$. Using the boundary condition (3.19), we get $A = 0$ so that

$$X(x) = X(0)\left[\frac{g}{\sigma} + \left(1 - \frac{g}{\sigma}\right)\cosh(\sqrt{\sigma}x)\right]. \tag{3.23}$$

Using the other boundary condition (3.20), we get the eigenvalue relationship

$$\frac{g}{\sigma} = \left(\frac{g}{\sigma} - 1\right)\cosh(\sqrt{\sigma}). \tag{3.24}$$

The above equation has no closed-form solution. However, in this particular example we can still find the stability region by finding values of g for which there are eigenvalues with zero real parts. First, we check if $\sigma = 0$ satisfies (3.24) for some values of g. Using the Taylor expansion for $\cosh(\sqrt{\sigma})$, we get

$$\frac{g}{\sigma} = \left(\frac{g}{\sigma} - 1\right)\left(1 + \frac{\sigma}{2} + O(\sigma^2)\right) = \frac{g}{\sigma} - 1 + \frac{g}{2} - \frac{\sigma}{2} + O(\sigma), \tag{3.25}$$

which gives $g = 2$ for $\sigma \to 0$. To show that there are no other eigenvalues on the imaginary axis, we set $\sigma = 2jy^2$, $y > 0$. Equation (3.24) then becomes

$$\cosh((j+1)y) = \frac{g}{g - 2jy^2},$$

$$\cos(y)\cosh(y) + j\sin(y)\sinh(y) = \frac{g^2 + 2jgy^2}{g^2 + 4y^4}.$$

Taking the absolute value, we get

$$\sinh(y)^2 + \cos(y)^2 = \frac{g^4 + 4g^2y^4}{(g^2 + 4y^4)^2}. \tag{3.26}$$

The only solution to this equation is $y = 0$, which can be seen by computing derivatives of both sides of (3.26):

$$\frac{d}{dy}(\sinh(y)^2 + \cos(y)^2) = \sinh(2y) - \sin(2y) > 0 \quad \text{for all } y > 0, \tag{3.27}$$

$$\frac{d}{dy}\frac{g^4 + 4g^2y^4}{(g^2 + 4y^4)^2} = -\frac{16g^2y^3}{(g^2 + 4y^4)^2} < 0 \quad \text{for all } y > 0. \tag{3.28}$$

Therefore, both sides of (3.26) start at the same point at $y = 0$, and for $y > 0$ the left-hand side monotonically grows while the right-hand side monotonically decays. We thus proved that the plant (3.18)–(3.20) is neutrally stable only when $g = 2$. Since we know that the plant is stable for $g = 0$, we conclude that $g < 2$ is the stability region and $g > 2$ is the instability region. ◇

3.2 Notes and References

The method of separation of variables is discussed in detail in classical PDE texts such as [40] and [187]. The exact solutions for many problems can be found in [33] and [140]. Transform methods for PDEs are studied extensively in [49].

Exercises

3.1. Consider the reaction-diffusion equation

$$u_t = u_{xx} + \lambda u$$

for $x \in (0, 1)$ with the initial condition $u_0(x) = u(x, 0)$ and boundary conditions

$$u_x(0) = 0\,,$$
$$u(1) = 0\,.$$

(a) Find the solution of this PDE.
(b) For what values of the parameter λ is this system unstable?

3.2. Consider the heat equation

$$u_t = u_{xx}$$

with Robin's boundary conditions

$$u_x(0) = -qu(0)\,,$$
$$u(1) = 0\,.$$

Find the range of values of the parameter q for which this system is unstable.

Chapter 4

Parabolic PDEs: Reaction-Advection-Diffusion and Other Equations

This is the most important chapter of this book. In this chapter we present the first designs of feedback laws for stabilization of PDEs using boundary control and introduce the method of *backstepping*.

As the reader shall learn throughout this book, there are many classes of PDEs—first and second order in time; first, second, third, fourth (and so on) order in space;[7] systems of coupled PDEs of various classes; PDEs interconnected with ODEs; real-valued and complex-valued PDEs;[8] and various other classes. Our introduction to boundary control, stabilization of PDEs, and the backstepping method is presented in this chapter for parabolic PDEs. There is no strong pedagogical reason why the introduction could not be done on some of the other classes of PDEs; however, parabolic PDEs are particularly convenient because they are both sufficiently simple and sufficiently general to serve as a launch pad from which one can easily extend the basic design tools to other classes of PDEs.

Parabolic PDEs are first order in time, which makes them more easily accessible to a reader with a background in ODEs, as opposed to second order in time PDEs such as wave equations, which, as we shall see in Chapter 7, have peculiarities that make defining the system "state," choosing a Lyapunov function, and adding damping to the system rather nonobvious.

This book deals exclusively with boundary control of PDEs. In-domain actuation of any kind (point actuation or distributed actuation) is not dealt with. The reasons for this are twofold. First, a considerable majority of problems in PDE control, particularly those involving fluids, can be actuated in a physically reasonable way only from the boundary. Second, the backstepping approach is particularly well suited for boundary control. Its earlier ODE applications provide a clue that it should be applicable also to many problems with in-domain actuation; however, at the moment, backstepping for PDEs is developed only for boundary control actuation.

[7] Respectively, we mean the transport equation, the heat and wave equations, the Korteweg–de Vries equation, and the Euler–Bernoulli beam and Kuramoto–Sivashinsky equations.

[8] Respectively, we mean the heat equation, the Schrödinger equation.

The main feature of backstepping is that it is capable of eliminating destabilizing effects ("forces" or "terms") that appear throughout the domain while the control is acting only from the boundary. This is at first a highly surprising result. However, as we shall see, the result is arrived at by following the standard approach in control of nonlinear ODEs, where nonlinearities "unmatched" by the control input can be dealt with using a combination of a diffeomorphic change of coordinates and feedback. We pursue a continuum equivalent of this approach and build a change of variables, which involves a Volterra integral operator that "absorbs" the destabilizing terms acting in the domain and allows the boundary control to completely eliminate their effect. The Volterra operator has a lower triangular structure similar to backstepping transformations for nonlinear ODEs–thus the name *backstepping*.

In Section 4.10 we shed more light on the connection between backstepping for ODEs and the infinite-dimensional backstepping that we pursue here.

4.1 Backstepping: The Main Idea

Let us start with the simplest unstable PDE, the reaction-diffusion equation:

$$u_t(x,t) = u_{xx}(x,t) + \lambda u(x,t)\,, \tag{4.1}$$

$$u(0,t) = 0\,, \tag{4.2}$$

$$u(1,t) = U(t)\,, \tag{4.3}$$

where λ is an arbitrary constant and $U(t)$ is the control input. The open-loop system (4.1), (4.2) (with $u(1,t) = 0$) is unstable with arbitrarily many unstable eigenvalues for sufficiently large λ.

Since the term λu is the source of instability, the natural objective for a boundary feedback is to "eliminate" this term. The main idea of the backstepping method is to use the coordinate transformation

$$w(x,t) = u(x,t) - \int_0^x k(x,y)u(y,t)\,dy \tag{4.4}$$

along with feedback control

$$u(1,t) = \int_0^1 k(1,y)u(y,t)\,dy \tag{4.5}$$

to transform the system (4.1), (4.2) into the *target* system

$$w_t(x,t) = w_{xx}(x,t)\,, \tag{4.6}$$

$$w(0,t) = 0\,, \tag{4.7}$$

$$w(1,t) = 0\,, \tag{4.8}$$

which is exponentially stable, as shown in Chapter 2. Note that the boundary conditions (4.2), (4.7) and (4.5), (4.8) are verified by (4.4) without any condition on $k(x,y)$.

The transformation (4.4) is called Volterra integral transformation. Its most characteristic feature is that the limits of integral range from 0 to x, not from 0 to 1. This makes it "spatially causal"; that is, for a given x the right-hand side of (4.4) depends only on the values of u in the interval $[0, x]$. Another important property of the Volterra transformation is that it is invertible, so that stability of the target system translates into stability of the closed-loop system consisting of the plant plus boundary feedback (see Section 4.5).

Our goal now is to find the function $k(x, y)$ (which we call the "gain kernel") that makes the plant (4.1), (4.2) with the controller (4.5) behave as the target system (4.6)–(4.8). It is not obvious at this point that such a function even exists.

4.2 Gain Kernel PDE

To find out what conditions $k(x, y)$ has to satisfy, we simply substitute the transformation (4.4) into the target system (4.6)–(4.8) and use the plant equations (4.1), (4.2). To do that, we need to differentiate the transformation (4.4) with respect to x and t, which is easy once we recall the Leibnitz differentiation rule:

$$\frac{d}{dx}\int_0^x f(x, y)\,dy = f(x, x) + \int_0^x f_x(x, y)\,dy\,.$$

We also introduce the following notation:

$$\begin{aligned}
k_x(x, x) &= \frac{\partial}{\partial x}k(x, y)|_{y=x}\,,\\
k_y(x, x) &= \frac{\partial}{\partial y}k(x, y)|_{y=x}\,,\\
\frac{d}{dx}k(x, x) &= k_x(x, x) + k_y(x, x).
\end{aligned}$$

Differentiating the transformation (4.4) with respect to x gives

$$\begin{aligned}
w_x(x) &= u_x(x) - k(x, x)u(x) - \int_0^x k_x(x, y)u(y)\,dy\,,\\
w_{xx}(x) &= u_{xx}(x) - \frac{d}{dx}(k(x, x)u(x)) - k_x(x, x)u(x) - \int_0^x k_{xx}(x, y)u(y)\,dy\,,\\
&= u_{xx}(x) - u(x)\frac{d}{dx}k(x, x) - k(x, x)u_x(x) - k_x(x, x)u(x)\\
&\quad - \int_0^x k_{xx}(x, y)u(y)\,dy\,. \qquad (4.9)
\end{aligned}$$

This expression for the second spatial derivative of $w(x)$ is going to be the same for different problems since no information about the specific plant and target system is used at this point.

Next, we differentiate the transformation (4.4) with respect to time:

$$
\begin{aligned}
w_t(x) &= u_t(x) - \int_0^x k(x,y)u_t(y)dy \\
&= u_{xx}(x) + \lambda u(x) - \int_0^x k(x,y)\left(u_{yy}(y) + \lambda u(y)\right) dy \\
&= u_{xx}(x) + \lambda u(x) - k(x,x)u_x(x) + k(x,0)u_x(0) \\
&\quad + \int_0^x k_y(x,y)u_y(y)\,dy - \int_0^x \lambda k(x,y)u(y)\,dy \quad \text{(integration by parts)} \\
&= u_{xx}(x) + \lambda u(x) - k(x,x)u_x(x) + k(x,0)u_x(0) + k_y(x,x)u(x) - k_y(x,0)u(0) \\
&\quad - \int_0^x k_{yy}(x,y)u(y)\,dy - \int_0^x \lambda k(x,y)u(y)\,dy \quad \text{(integration by parts)}. \qquad (4.10)
\end{aligned}
$$

Subtracting (4.9) from (4.10), we get

$$
\begin{aligned}
w_t - w_{xx} = &\left[\lambda + 2\frac{d}{dx}k(x,x)\right] u(x) + k(x,0)u_x(0) \\
&+ \int_0^x \left(k_{xx}(x,y) - k_{yy}(x,y) - \lambda k(x,y)\right) u(y)\,dy\,. \qquad (4.11)
\end{aligned}
$$

For the right-hand side to be zero for all u, the following three conditions have to be satisfied:

$$k_{xx}(x,y) - k_{yy}(x,y) - \lambda k(x,y) = 0\,, \qquad (4.12)$$

$$k(x,0) = 0\,, \qquad (4.13)$$

$$\lambda + 2\frac{d}{dx}k(x,x) = 0. \qquad (4.14)$$

We can simplify (4.14) by integrating it with respect to x and noting from (4.13) that $k(0,0) = 0$, which gives us the following:

$$
\boxed{
\begin{aligned}
k_{xx}(x,y) - k_{yy}(x,y) &= \lambda k(x,y)\,, \\
k(x,0) &= 0\,, \\
k(x,x) &= -\frac{\lambda}{2}x\,.
\end{aligned}
} \qquad (4.15)
$$

It turns out that these three conditions are compatible and in fact form a well-posed PDE. This PDE is of hyperbolic type: one can think of it as a wave equation with an extra term λk (x plays the role of time). In quantum physics such PDEs are called Klein–Gordon PDEs. The domain of this PDE is a triangle $0 \le y \le x \le 1$ and is shown in Figure 4.1. The boundary conditions are prescribed on two sides of the triangle and the third side (after solving for $k(x,y)$) gives us the control gain $k(1,y)$.

In the next two sections we prove that the PDE (4.15) has a unique twice continuously differentiable solution.

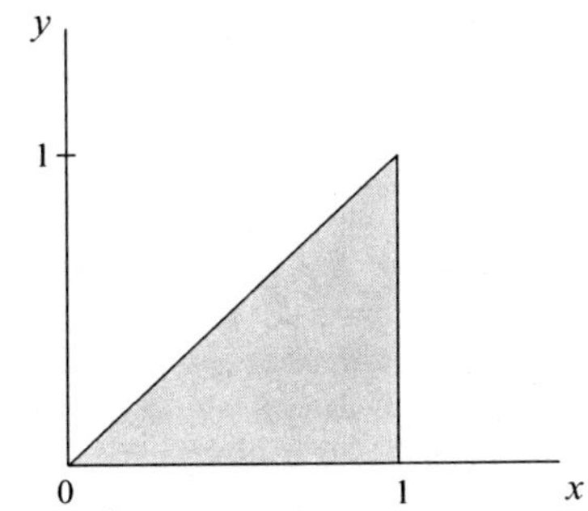

Figure 4.1. *Domain of the gain kernel PDE.*

4.3 Converting the Gain Kernel PDE into an Integral Equation

To find a solution of the PDE (4.15) we first convert it into an integral equation. Introducing the change of variables

$$\xi = x + y, \qquad \eta = x - y, \tag{4.16}$$

we have

$$\begin{aligned} k(x,y) &= G(\xi,\eta)\,,\\ k_x &= G_\xi + G_\eta\,,\\ k_{xx} &= G_{\xi\xi} + 2G_{\xi\eta} + G_{\eta\eta}\,,\\ k_y &= G_\xi - G_\eta\,,\\ k_{yy} &= G_{\xi\xi} - 2G_{\xi\eta} + G_{\eta\eta}\,. \end{aligned}$$

Thus, the gain kernel PDE becomes

$$G_{\xi\eta}(\xi,\eta) = \frac{\lambda}{4}G(\xi,\eta)\,, \tag{4.17}$$

$$G(\xi,\xi) = 0\,, \tag{4.18}$$

$$G(\xi,0) = -\frac{\lambda}{4}\xi\,. \tag{4.19}$$

Integrating (4.17) with respect to η from 0 to η, we get

$$G_\xi(\xi,\eta) = G_\xi(\xi,0) + \int_0^\eta \frac{\lambda}{4}G(\xi,s)\,ds = -\frac{\lambda}{4} + \int_0^\eta \frac{\lambda}{4}G(\xi,s)\,ds\,. \tag{4.20}$$

Next, we integrate (4.20) with respect to ξ from η to ξ to get

$$\begin{aligned} G(\xi,\eta) &= G(\eta,\eta) - \frac{\lambda}{4}(\xi-\eta) + \frac{\lambda}{4}\int_\eta^\xi \int_0^\eta G(\tau,s)\,ds\,d\tau \\ &= -\frac{\lambda}{4}(\xi-\eta) + \frac{\lambda}{4}\int_\eta^\xi \int_0^\eta G(\tau,s)\,ds\,d\tau\,. \end{aligned} \tag{4.21}$$

We obtained the integral equation, which is equivalent to PDE (4.15) in the sense that every solution of (4.15) is a solution of (4.21). The point of converting the PDE into the integral equation is that the latter is easier to analyze with a special tool, which consider next.

4.4 Method of Successive Approximations

The method of successive approximations is conceptually simple: start with an initial guess for a solution of the integral equation, substitute it into the right-hand side of the equation, then use the obtained expression as the next guess in the integral equation and repeat the process. Eventually this process results in a solution of the integral equation.

Let us start with an initial guess

$$G^0(\xi, \eta) = 0 \tag{4.22}$$

and set up the recursive formula for (4.21) as follows:

$$G^{n+1}(\xi, \eta) = -\frac{\lambda}{4}(\xi - \eta) + \frac{\lambda}{4}\int_\eta^\xi \int_0^\eta G^n(\tau, s)ds\, d\tau\,. \tag{4.23}$$

If this converges, we can write the solution $G(\xi, \eta)$ as

$$G(\xi, \eta) = \lim_{n\to\infty} G^n(\xi, \eta)\,. \tag{4.24}$$

Let us denote the difference between two consecutive terms as

$$\Delta G^n(\xi, \eta) = G^{n+1}(\xi, \eta) - G^n(\xi, \eta)\,. \tag{4.25}$$

Then

$$\Delta G^{n+1}(\xi, \eta) = \frac{\lambda}{4}\int_\eta^\xi \int_0^\eta \Delta G^n(\tau, s)ds\, d\tau \tag{4.26}$$

and (4.24) can be alternatively written as

$$G(\xi, \eta) = \sum_{n=0}^{\infty} \Delta G^n(\xi, \eta)\,. \tag{4.27}$$

Computing ΔG^n from (4.26) starting with

$$\Delta G^0 = G^1(\xi, \eta) = -\frac{\lambda}{4}(\xi - \eta)\,, \tag{4.28}$$

we can observe the pattern which leads to the following formula:

$$\Delta G^n(\xi, \eta) = -\frac{(\xi - \eta)\xi^n\eta^n}{n!(n+1)!}\left(\frac{\lambda}{4}\right)^{n+1}\,. \tag{4.29}$$

This formula can be verified by induction. The solution to the integral equation is given by

$$G(\xi, \eta) = -\sum_{n=0}^{\infty} \frac{(\xi - \eta)\xi^n\eta^n}{n!(n+1)!}\left(\frac{\lambda}{4}\right)^{n+1}\,. \tag{4.30}$$

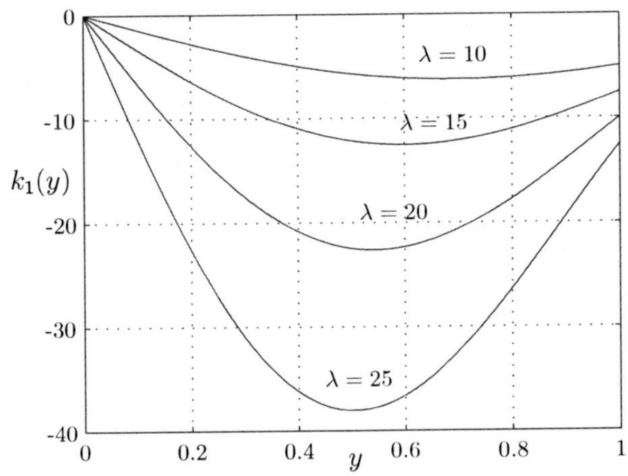

Figure 4.2. *Control gain $k(1, y)$ for different values of λ.*

To compute the series (4.30), note from the appendix that a first-order modified Bessel function of the first kind can be represented as

$$I_1(x) = \sum_{n=0}^{\infty} \frac{\left(\frac{x}{2}\right)^{2n+1}}{n!(n+1)!}. \tag{4.31}$$

Comparing this expression with (4.30), we obtain

$$G(\xi, \eta) = -\frac{\lambda}{2}(\xi - \eta)\frac{I_1(\sqrt{\lambda \xi \eta})}{\sqrt{\lambda \xi \eta}} \tag{4.32}$$

or, returning to the original x, y variables,

$$\boxed{k(x, y) = -\lambda y \frac{I_1\left(\sqrt{\lambda(x^2 - y^2)}\right)}{\sqrt{\lambda(x^2 - y^2)}}}. \tag{4.33}$$

In Figure 4.2 the control gain $k(1, y)$ is shown for different values of λ. Obviously, as λ gets larger, the plant becomes more unstable, which requires more control effort. Low control gain near the boundaries is also logical: near $x = 0$ the state is small even without control because of the boundary condition $u(0) = 0$, and near $x = 1$ the control has the most authority.

4.5 Inverse Transformation

To complete the design we need to establish that stability of the target system (4.6)–(4.8) implies stability of the closed-loop plant (4.1), (4.2), (4.5). In other words, we need to show that the transformation (4.4) is invertible.

Let us write an inverse transformation in the form

$$u(x) = w(x) + \int_0^x l(x, y) w(y)\, dy\,, \tag{4.34}$$

where $l(x, y)$ is the transformation kernel.

Given the direct transformation (4.4) and the inverse transformation (4.34), the kernels $k(x, y)$ and $l(x, y)$ satisfy

$$\boxed{l(x, y) = k(x, y) + \int_y^x k(x, \xi) l(\xi, y)\, d\xi\,.} \tag{4.35}$$

Proof of **(4.35).** First, let us recall from calculus the following formula for changing the order of integration:

$$\int_0^x \int_0^y f(x, y, \xi)\, d\xi\, dy = \int_0^x \int_\xi^x f(x, y, \xi)\, dy\, d\xi. \tag{4.36}$$

Substituting (4.34) into (4.4), we get

$$\begin{aligned} w(x) &= w(x) + \int_0^x l(x, y) w(y) dy - \int_0^x k(x, y) \left[w(y) + \int_0^y l(y, \xi) w(\xi) d\xi \right] dy \\ &= w(x) + \int_0^x l(x, y) w(y) dy - \int_0^x k(x, y) w(y) dy - \int_0^x \int_0^y k(x, y) l(y, \xi) w(\xi) d\xi dy\,, \\ 0 &= \int_0^x w(y) \left[l(x, y) - k(x, y) - \int_y^x k(x, \xi) l(\xi, y)\, d\xi \right] dy\,. \end{aligned}$$

Since the last line has to hold for all $w(y)$, we get the relationship (4.35). □

The formula (4.35) is general (it does not depend on the plant and the target system) but is not very helpful in actually finding $l(x, y)$ from $k(x, y)$. Instead, we follow the same approach that led us to the kernel PDE for $k(x, y)$: we differentiate the inverse transformation (4.34) with respect to x and t and use the plant and the target system to obtain the PDE for $l(x, y)$.

Differentiating (4.34) with respect to time, we get

$$\begin{aligned} u_t(x) &= w_t(x) + \int_0^x l(x, y) w_t(y)\, dy \\ &= w_{xx}(x) + l(x, x) w_x(x) - l(x, 0) w_x(0) - l_y(x, x) w(x) \\ &\quad + \int_0^x l_{yy}(x, y) w(y)\, dy\,, \end{aligned} \tag{4.37}$$

and differentiating twice with respect to x gives

$$\begin{aligned} u_{xx}(x) &= w_{xx}(x) + l_x(x, x) w(x) + w(x) \frac{d}{dx} l(x, x) + l(x, x) w_x(x) \\ &\quad + \int_0^x l_{xx}(x, y) w(y)\, dy\,. \end{aligned} \tag{4.38}$$

Subtracting (4.38) from (4.37), we get

$$\lambda w(x) + \lambda \int_0^x l(x,y)w(y)dy = -2w(x)\frac{d}{dx}l(x,x) - l(x,0)w_x(0) + \int_0^x (l_{yy}(x,y) - l_{xx}(x,y))w(y)\,dy\,,$$

which gives the following conditions on $l(x,y)$:

$$l_{xx}(x,y) - l_{yy}(x,y) = -\lambda l(x,y)\,, \tag{4.39}$$

$$l(x,0) = 0\,, \tag{4.40}$$

$$l(x,x) = -\frac{\lambda}{2}x\,. \tag{4.41}$$

Comparing this PDE with the PDE (4.15) for $k(x,y)$, we see that

$$l(x,y;\lambda) = -k(x,y;-\lambda)\,. \tag{4.42}$$

From (4.33) we have

$$l(x,y) = -\lambda y \frac{I_1\left(\sqrt{-\lambda(x^2-y^2)}\right)}{\sqrt{-\lambda(x^2-y^2)}} - \lambda y \frac{I_1\left(j\sqrt{\lambda(x^2-y^2)}\right)}{j\sqrt{\lambda(x^2-y^2)}}$$

or, using the properties of I_1 (see the appendix),

$$\boxed{l(x,y) = -\lambda y \frac{J_1\left(\sqrt{\lambda(x^2-y^2)}\right)}{\sqrt{\lambda(x^2-y^2)}}}\,. \tag{4.43}$$

A summary of the control design for the plant (4.1), (4.2) is presented in Table 4.1.

Note that, since the solutions to the target system (4.6)–(4.8) can be found explicitly and the direct and inverse transformations (4.4), (4.34) are explicit as well, it is possible to derive the explicit solution to the closed-loop system; see Exercises 4.3 and 4.4 for an example of how this is done.

In Figures 4.3 and 4.4 simulation results for the scheme (4.44)–(4.46) are presented for the case $\lambda = 20$. The plant has one unstable eigenvalue $20 - \pi^2 \approx 10$. In the top graph of Figure 4.3 one can see that the state of the uncontrolled plant quickly grows. Note that the initial condition is rapidly smoothed out even though the plant is unstable, and then the state takes the shape of the eigenfunction $\sin(\pi x)$ which corresponds to the unstable eigenvalue. The bottom graph of Figure 4.3 shows the response of the controlled plant. The instability is quickly suppressed and the state converges to the zero equilibrium. The control is shown in Figure 4.4.

In the next example, we consider a reaction-diffusion plant with a Neumann, rather than a Dirichlet, boundary condition at the uncontrolled end, and once again go through all the design steps.

Table 4.1. *Summary of control design for the reaction-diffusion equation.*

Plant:

$$u_t = u_{xx} + \lambda u \,, \tag{4.44}$$
$$u(0) = 0 \,. \tag{4.45}$$

Controller:

$$u(1) = -\int_0^1 y\lambda \frac{I_1\left(\sqrt{\lambda(1-y^2)}\right)}{\sqrt{\lambda(1-y^2)}} u(y)\, dy \,. \tag{4.46}$$

Transformation:

$$w(x) = u(x) + \int_0^x \lambda y \frac{I_1\left(\sqrt{\lambda(x^2-y^2)}\right)}{\sqrt{\lambda(x^2-y^2)}} u(y)\, dy \,, \tag{4.47}$$

$$u(x) = w(x) - \int_0^x \lambda y \frac{J_1\left(\sqrt{\lambda(x^2-y^2)}\right)}{\sqrt{\lambda(x^2-y^2)}} w(y)\, dy \,. \tag{4.48}$$

Target system:

$$w_t = w_{xx} \,, \tag{4.49}$$
$$w(0) = 0 \,, \tag{4.50}$$
$$w(1) = 0 \,. \tag{4.51}$$

Example 4.1 Consider the plant

$$u_t = u_{xx} + \lambda u \,, \tag{4.52}$$
$$u_x(0) = 0 \,, \tag{4.53}$$
$$u(1) = U(t) \,. \tag{4.54}$$

We use the transformation

$$w(x) = u(x) - \int_0^x k(x,y) u(y)\, dy \tag{4.55}$$

to map this plant into the target system

$$w_t = w_{xx} \,, \tag{4.56}$$
$$w_x(0) = 0 \,, \tag{4.57}$$
$$w(1) = 0 \,. \tag{4.58}$$

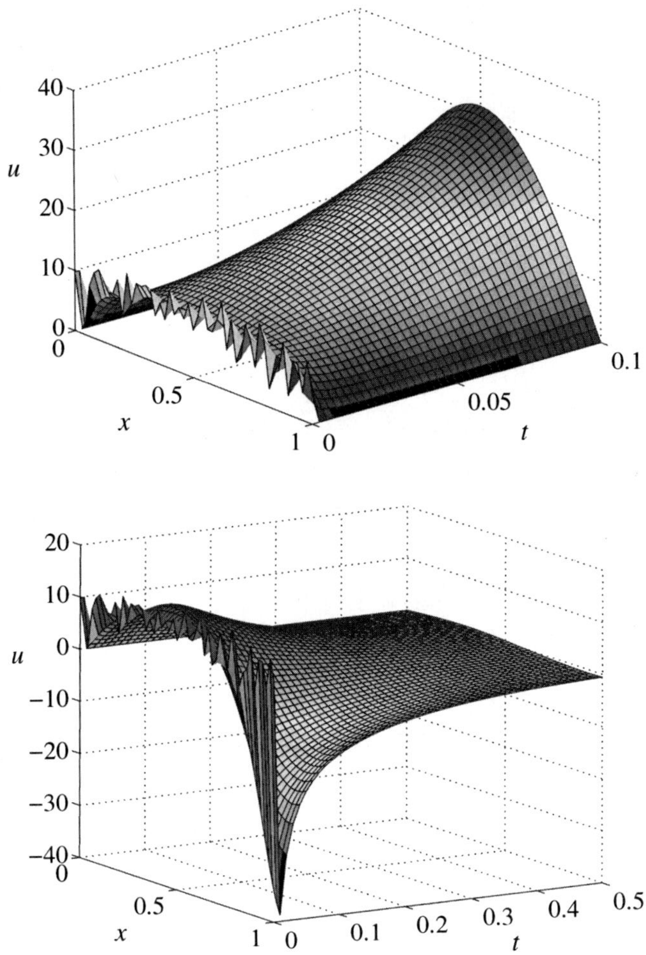

Figure 4.3. *Simulation results for reaction-diffusion plant* (4.44), (4.45). *Top: open-loop response. Bottom: closed-loop response with controller* (4.46) *implemented.*

Differentiation of the transformation (4.55) with respect to x gives (4.9) (which does not depend on the particular plant). Differentiating (4.55) with respect to time, we get

$$
\begin{aligned}
w_t(x) &= u_t(x) - \int_0^x k(x,y)u_t(y)\,dy \\
&= u_{xx}(x) + \lambda u(x) - \int_0^x k(x,y)[u_{yy}(y) + \lambda u(y)]\,dy \\
&= u_{xx}(x) + \lambda u(x) - k(x,x)u_x(x) + k(x,0)u_x(0) \\
&\quad + \int_0^x k_y(x,y)u_y(y)dy - \int_0^x \lambda k(x,y)u(y)\,dy \quad \text{(integration by parts)} \\
&= u_{xx}(x) + \lambda u(x) - k(x,x)u_x(x) + k_y(x,x)u(x) - k_y(x,0)u(0) \\
&\quad - \int_0^x k_{yy}(x,y)u(y)\,dy - \int_0^x \lambda k(x,y)u(y)\,dy \quad \text{(integration by parts).}
\end{aligned}
\tag{4.59}
$$

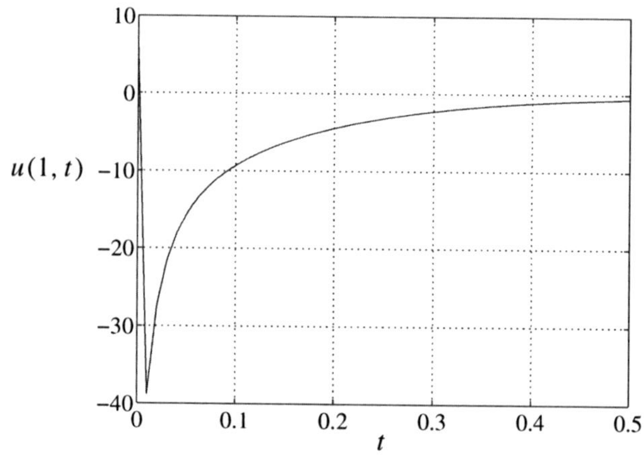

Figure 4.4. *The control* (4.46) *for reaction-diffusion plant* (4.44), (4.45).

Subtracting (4.9) from (4.59), we get

$$w_t - w_{xx} = \left[\lambda + 2\frac{d}{dx}k(x,x)\right]u(x) - k_y(x,0)u(0) + \int_0^x \big(k_{xx}(x,y) - k_{yy}(x,y) - \lambda k(x,y)\big)\,u(y)\,dy\,. \tag{4.60}$$

For the right-hand side of this equation to be zero for all $u(x)$, the following three conditions must be satisfied:

$$k_{xx}(x,y) - k_{yy}(x,y) - \lambda k(x,y) = 0\,, \tag{4.61}$$

$$k_y(x,0) = 0\,, \tag{4.62}$$

$$\lambda + 2\frac{d}{dx}k(x,x) = 0\,. \tag{4.63}$$

Integrating (4.63) with respect to x gives $k(x,x) = -\lambda/2x + k(0,0)$, where $k(0,0)$ is obtained using the boundary condition (4.57),

$$w_x(0) = u_x(0) + k(0,0)u(0) = 0\,,$$

so that $k(0,0) = 0$. The gain kernel PDE is thus

$$k_{xx}(x,y) - k_{yy}(x,y) = \lambda k(x,y)\,, \tag{4.64}$$

$$k_y(x,0) = 0\,, \tag{4.65}$$

$$k(x,x) = -\frac{\lambda}{2}x\,. \tag{4.66}$$

Note that this PDE is very similar to (4.15); the only difference is in the boundary condition at $y = 0$. The solution to the PDE (4.64)–(4.66) is obtained through a summation of successive approximation series, similarly to the way it was obtained for the PDE (4.15):

$$k(x, y) = -\lambda x \frac{I_1\left(\sqrt{\lambda(x^2 - y^2)}\right)}{\sqrt{\lambda(x^2 - y^2)}}. \tag{4.67}$$

Thus, the controller is given by

$$u(1) = -\int_0^1 \lambda \frac{I_1\left(\sqrt{\lambda(1 - y^2)}\right)}{\sqrt{\lambda(1 - y^2)}} u(y)\, dy. \tag{4.68}$$

◇

4.6 Neumann Actuation

So far we considered only the case of Dirichlet actuation (where $u(1)$ is controlled), which is usually the case in fluid problems where the velocity is controlled using microjets. In problems with thermal and chemically reacting dynamics, the natural choice is the Neumann actuation (where $u_x(1)$, or heat flux, is controlled). The Neumann controllers are obtained using the same exact transformation (4.4) as in the case of the Dirichlet actuation, but with the appropriate change in the boundary condition of the target system (from Dirichlet to Neumann).

To illustrate the design procedure, consider the plant (4.1), (4.2) but with $u_x(1)$ actuated:

$$u_t = u_{xx} + \lambda u, \tag{4.69}$$

$$u(0) = 0, \tag{4.70}$$

$$u_x(1) = U(t). \tag{4.71}$$

We use the same transformation (4.4), (4.33) as we used in the case of Dirichlet actuation. To obtain the control $u_x(1)$, we need to differentiate (4.4) with respect to x:

$$w_x(x) = u_x(x) - k(x, x)u(x) - \int_0^x k_x(x, y)u(y)\, dy$$

and set $x = 1$. It is clear now that the target system has to have the Neumann boundary condition at $x = 1$:

$$w_t = w_{xx}, \tag{4.72}$$

$$w(0) = 0, \tag{4.73}$$

$$w_x(1) = 0, \tag{4.74}$$

which gives the controller

$$u_x(1) = k(1, 1)u(1) + \int_0^1 k_x(1, y)u(y)\, dy. \tag{4.75}$$

All that remains is to derive the expression for k_x from (4.33) using the properties of Bessel functions given in the appendix:

$$k_x(x, y) = -\lambda y x \frac{I_2\left(\sqrt{\lambda(x^2 - y^2)}\right)}{x^2 - y^2}.$$

Finally, the controller is

$$u_x(1) = -\frac{\lambda}{2}u(1) - \int_0^1 \lambda y \frac{I_2\left(\sqrt{\lambda(1 - y^2)}\right)}{1 - y^2} u(y)\, dy. \tag{4.76}$$

4.7 Reaction-Advection-Diffusion Equation

Consider the reaction-advection-diffusion equation

$$u_t = \varepsilon u_{xx} + b u_x + \lambda u\,, \tag{4.77}$$

$$u(0) = 0\,, \tag{4.78}$$

$$u(1) = U(t)\,. \tag{4.79}$$

It is possible to design an explicit controller for this system based on the formula (4.33). First, we eliminate the advection term u_x from this equation with the following change of variables:

$$v(x) = u(x) e^{\frac{b}{2\varepsilon}x}\,. \tag{4.80}$$

Taking the temporal and spatial derivatives, we get

$$u_t(x) = v_t(x) e^{-\frac{b}{2\varepsilon}x}\,,$$

$$u_x(x) = v_x(x) e^{-\frac{b}{2\varepsilon}x} - \frac{b}{2\varepsilon} v(x) e^{-\frac{b}{2\varepsilon}x}\,,$$

$$u_{xx}(x) = v_{xx}(x) e^{-\frac{b}{2\varepsilon}x} - \frac{b}{\varepsilon} v_x(x) e^{-\frac{b}{2\varepsilon}x} + \frac{b^2}{4\varepsilon^2} v(x) e^{-\frac{b}{2\varepsilon}x}\,,$$

and the equation (4.77) becomes

$$v_t e^{-\frac{b}{2\varepsilon}x} = \left(\varepsilon v_{xx} - b v_x + \frac{b^2}{4\varepsilon} v + b v_x - \frac{b^2}{2\varepsilon} v + \lambda v\right) e^{-\frac{b}{2\varepsilon}x}\,, \tag{4.81}$$

which gives

$$v_t = \varepsilon v_{xx} + \left(\lambda - \frac{b^2}{4\varepsilon}\right) v\,, \tag{4.82}$$

$$v(0) = 0\,, \tag{4.83}$$

$$v(1) = u(1) e^{\frac{b}{2\varepsilon}} = \text{control}. \tag{4.84}$$

Now the transformation

$$w(x) = v(x) - \int_0^x k(x, y) v(y) dy \tag{4.85}$$

leads to the target system

$$w_t = \varepsilon w_{xx} - cw\,, \tag{4.86}$$

$$w(0) = 0\,, \tag{4.87}$$

$$w(1) = 0\,. \tag{4.88}$$

Here the constant c is a design parameter that sets the decay rate of the closed-loop system. It should satisfy the following stability condition:

$$c \geq \max\left\{\frac{b^2}{4\varepsilon} - \lambda, 0\right\}.$$

The max is used to prevent unnecessary spending of control effort when the plant is stable. The gain kernel $k(x, y)$ can be shown to satisfy the following PDE:

$$\varepsilon k_{xx}(x, y) - \varepsilon k_{yy}(x, y) = \left(\lambda - \frac{b^2}{4\varepsilon} + c\right) k(x, y)\,, \tag{4.89}$$

$$k(x, 0) = 0\,, \tag{4.90}$$

$$k(x, x) = -\frac{x}{2\varepsilon}\left(\lambda - \frac{b^2}{4\varepsilon} + c\right). \tag{4.91}$$

This equation is exactly the same as (4.15), just with a constant different from λ and denoted by λ_0:

$$\lambda_0 = \frac{1}{\varepsilon}\left(\lambda - \frac{b^2}{4\varepsilon} + c\right). \tag{4.92}$$

Therefore the solution to (4.89)–(4.91) is given by

$$k(x, y) = -\lambda_0 y \frac{I_1\left(\sqrt{\lambda_0(x^2 - y^2)}\right)}{\sqrt{\lambda_0(x^2 - y^2)}} \tag{4.93}$$

and the controller is

$$u(1) = \int_0^1 \mathrm{e}^{-\frac{b}{2\varepsilon}(1-y)} \lambda_0 y \frac{I_1\left(\sqrt{\lambda_0(1 - y^2)}\right)}{\sqrt{\lambda_0(1 - y^2)}} u(y)\, dy\,. \tag{4.94}$$

Let us examine the effect of the advection term bu_x in (4.77) on open-loop stability and on the size of the control gain. From (4.82) we see that the advection term has a beneficial effect on open-loop stability, irrespective of the sign of the advection coefficient b. However, the effect of b on the gain function in the control law in (4.94) is "sign-sensitive." Negative values of b demand a much higher control effort than positive values of b. Interestingly, negative values of b refer to the situation where the state disturbances advect towards the actuator at $x = 1$, whereas the "easier" case of positive b refers to the case where the state disturbances advect away from the actuator at $x = 1$ and towards the Dirichlet boundary condition (4.78) at $x = 0$.

4.8 Reaction-Advection-Diffusion Systems with Spatially Varying Coefficients

The most general one-dimensional linear reaction-advection-diffusion PDE has the form

$$u_t = \varepsilon(x)u_{xx} + b(x)u_x + \lambda(x)u\,, \tag{4.95}$$

$$u_x(0) = -qu(0)\,, \tag{4.96}$$

where $u(1)$ is actuated. This equation describes a variety of systems with thermal, fluid, and chemically reacting dynamics. The spatially varying coefficients come from applications with nonhomogenous materials and unusually shaped domains and can also arise from linearization. Also note the mixed boundary condition at $x = 0$.

Using a so-called gauge transformation, it is possible to convert this system into one with constant diffusion and zero advection terms. Consider a coordinate change

$$z = \sqrt{\varepsilon_0}\int_0^x \frac{ds}{\sqrt{\varepsilon(s)}}, \quad \text{where } \varepsilon_0 = \left(\int_0^1 \frac{ds}{\sqrt{\varepsilon(s)}}\right)^{-2}, \tag{4.97}$$

and a change of the state variable

$$v(z) = \varepsilon^{-1/4}(x)u(x)e^{\int_0^x \frac{b(s)}{2\varepsilon(s)}\,ds}\,. \tag{4.98}$$

Then one can show that the new state v satisfies the following PDE:

$$v_t(z,t) = \varepsilon_0 v_{zz}(z,t) + \lambda_0(z)v(z,t)\,, \tag{4.99}$$

$$v_z(0,t) = -q_0 v(0,t), \tag{4.100}$$

where

$$\varepsilon_0 = \left(\int_0^1 \frac{ds}{\sqrt{\varepsilon(s)}}\right)^{-2} \tag{4.101}$$

$$\lambda_0(z) = \lambda(x) + \frac{\varepsilon''(x)}{4} - \frac{b'(x)}{2} - \frac{3}{16}\frac{(\varepsilon'(x))^2}{\varepsilon(x)} + \frac{1}{2}\frac{b(x)\varepsilon'(x)}{\varepsilon(x)} - \frac{1}{4}\frac{b^2(x)}{\varepsilon(x)} \tag{4.102}$$

$$q_0 = q\sqrt{\frac{\varepsilon(0)}{\varepsilon_0}} - \frac{b(0)}{2\sqrt{\varepsilon_0\varepsilon(0)}} - \frac{\varepsilon'(0)}{4\sqrt{\varepsilon_0\varepsilon(0)}}. \tag{4.103}$$

We use the transformation (4.4) to map the modified plant into the target system

$$w_t = \varepsilon_0 w_{zz} - cw\,, \tag{4.104}$$

$$w_z(0) = 0\,, \tag{4.105}$$

$$w(1) = 0\,. \tag{4.106}$$

Here c is a design parameter that determines the decay rate of the closed-loop system. The transformation kernel is found by solving the PDE

$$k_{zz}(z,y) - k_{yy}(z,y) = \frac{\lambda_0(y)+c}{\varepsilon_0}k(z,y)\,, \tag{4.107}$$

$$k_y(z,0) = -q_0 k(z,0)\,, \tag{4.108}$$

$$k(z,z) = -q_0 - \frac{1}{2\varepsilon_0}\int_0^z (\lambda_0(y)+c)\,dy\,. \tag{4.109}$$

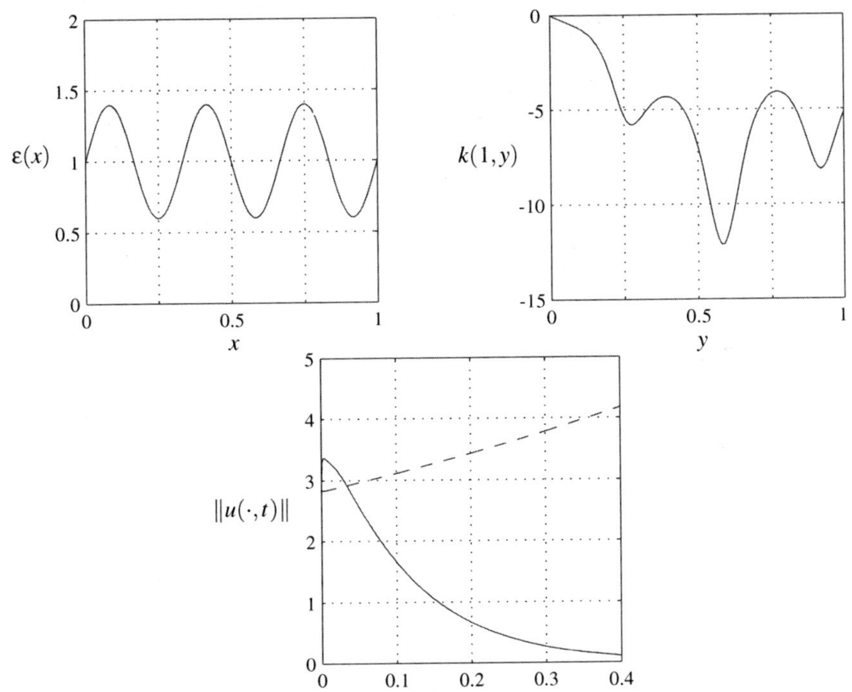

Figure 4.5. *Simulation results for* (4.95)–(4.96) *with controller* (4.111). *Top left:* $\varepsilon(x)$. *Top right: The gain kernel. Bottom: The* L_2*-norm of open-loop (dashed) and closed-loop (solid) response.*

This PDE can be shown to be well posed but, unlike the gain kernel equations for the plants we considered before, it cannot be solved in closed form. However, one can solve it either symbolically, using the recursive procedure similar to the one given in Section 4.4, or numerically with finite difference schemes developed for Klein–Gordon-type PDEs, which is substantially easier than solving an operator Riccati equation arising when one pursues optimality.

Since the controller for the v-system is given by

$$v(1) = \int_0^1 k(1, y)v(y)\,dy\,, \tag{4.110}$$

using (4.97) and (4.98) we obtain the controller for the original u-plant:

$$u(1) = \int_0^1 \frac{\varepsilon^{1/4}(1)\sqrt{\varepsilon_0}}{\varepsilon^{3/4}(y)} \mathrm{e}^{-\int_y^1 \frac{b(s)}{2\varepsilon(s)}ds} k\left(\int_0^1 \sqrt{\frac{\varepsilon_0}{\varepsilon(s)}}ds, \int_0^y \sqrt{\frac{\varepsilon_0}{\varepsilon(s)}}ds\right) u(y)\,dy\,. \tag{4.111}$$

The results of a closed-loop simulation are presented in Figure 4.5 for $\varepsilon(x) = 1 + 0.4\sin(6\pi x)$, $b \equiv 0$, and $\lambda = 10$.

4.9 Other Spatially Causal Plants

In this section we design backstepping controllers for plants that are not as common as reaction-advection-diffusion systems but are important nevertheless and have the "spatially causal" structure, which makes them tractable by our method. Consider the plant

$$u_t = u_{xx} + g(x)u(0) + \int_0^x f(x, y)u(y)dy\,, \tag{4.112}$$

$$u_x(0) = 0\,, \tag{4.113}$$

where $u(1)$ is actuated. This equation is partly motivated by the model of unstable burning in solid propellant rockets [28]. This plant also often appears as a part of the control design for more complicated systems (see, e.g., our Chapter 8 or [174]). Without the derivation, we present the PDE for the gain kernel:

$$k_{xx} - k_{yy} = -f(x, y) + \int_y^x k(x, \xi)f(\xi, y)\,d\xi\,, \tag{4.114}$$

$$k_y(x, 0) = g(x) - \int_0^x k(x, y)g(y)\,dy\,, \tag{4.115}$$

$$k(x, x) = 0\,. \tag{4.116}$$

In general, this equation has to be solved numerically. We consider one case when the solution can be obtained explicitly. Let $f \equiv 0$; then (4.114) becomes

$$k_{xx} - k_{yy} = 0\,, \tag{4.117}$$

which has a general solution of the form

$$k(x, y) = \phi(x - y) + \psi(x + y). \tag{4.118}$$

From the boundary condition (4.116) we get

$$\phi(0) + \psi(2x) = 0\,, \tag{4.119}$$

which means that, without loss of generality, we can set $\psi \equiv 0$ and $\phi(0) = 0$. Therefore, $k(x, y) = \phi(x - y)$. Substituting this expression into the boundary condition (4.115), we get

$$\phi'(x) = g(x) - \int_0^x \phi(x - y)g(y)\,dy\,. \tag{4.120}$$

Applying to this equation the Laplace transform with respect to x, we obtain

$$\begin{aligned} -s\phi(s) + \phi(0) &= g(s) - \phi(s)g(s)\,, \\ \phi(s) &= \frac{g(s)}{g(s) - s}\,. \end{aligned} \tag{4.121}$$

Thus, for any function $g(x)$ one can obtain $k(x, y)$ in closed form.

Example 4.2 Let $g(x) = g$. Then (4.121) gives

$$g(s) = \frac{g}{s},$$

and $\phi(s)$ becomes

$$\phi(s) = \frac{g}{g - s^2} = -\sqrt{g}\frac{\sqrt{g}}{s^2 - g}.$$

This gives

$$\phi(z) = -\sqrt{g}\sinh(\sqrt{g}z)$$

and

$$k(x, y) = -\sqrt{g}\sinh(\sqrt{g}(x - y)).$$

Therefore, for the plant

$$u_t = u_{xx} + gu(0),$$
$$u_x(0) = 0$$

the stabilizing controller is given by

$$u(1) = -\int_0^1 \sqrt{g}\sinh(\sqrt{g}(1 - y))u(y)\,dy. \quad \diamond$$

4.10 Comparison with ODE Backstepping

Before we close this chapter, in which we have introduced the method of *backstepping*, let us discuss the rationale behind the method's name. The method and its name originated in the early 1990s [80, 97] and is linked to problems of stabilization of nonlinear ODE systems. Consider the following three-state nonlinear system:

$$\dot{y}_1 = y_2 + y_1^3, \tag{4.122}$$
$$\dot{y}_2 = y_3 + y_2^3, \tag{4.123}$$
$$\dot{y}_3 = u + y_3^3. \tag{4.124}$$

Since the control input u is in only the last line of (4.124), it is helpful to view it as boundary control. The nonlinear terms y_1^3, y_2^3, y_3^3 can be viewed as nonlinear "reaction" terms and they are clearly destabilizing because, for $u = 0$, the overall system is a "cascade" of three unstable subsystems of the form $\dot{y}_i = y_i^3$ (the open-loop system exhibits a finite-time escape instability). The control u can cancel the "matched" term y_3^3 in (4.124) but cannot cancel directly the unmatched terms y_1^3 and y_2^3 in (4.122) and (4.124). To achieve the cancellation of all three destabilizing y_i^3-terms, a backstepping change of variables is constructed recursively,

$$z_1 = y_1, \tag{4.125}$$
$$z_2 = y_2 + y_1^3 + cy_1, \tag{4.126}$$
$$z_3 = y_3 + y_2^3 + (3y_1^2 + 2c)y_2 + 3y_1^5 + 2cy_1^3 + (c^2 + 1)y_1, \tag{4.127}$$

along with the control law

$$u = -cz_3 - z_2 - y_3^3 - (3y_2^2 + 3y_1^2 + 2c)(y_3 + y_2^3), \\ -(6y_1y_2 + 15y_1^4 + 6cy_1^2 + c^2 + 1)(y_2 + y_1^3), \tag{4.128}$$

which convert the system (4.122)–(4.124) into

$$\dot{z}_1 = z_2 - cz_1, \tag{4.129}$$

$$\dot{z}_2 = -z_1 + z_3 - cz_2, \tag{4.130}$$

$$\dot{z}_3 = -z_2 - cz_3, \tag{4.131}$$

where the control parameter c should be chosen as positive. The system (4.129)–(4.131), which can also be written as

$$\dot{z} = Az, \tag{4.132}$$

where

$$A = \begin{bmatrix} -c & 1 & 0 \\ -1 & -c & 1 \\ 0 & -1 & -c \end{bmatrix}, \tag{4.133}$$

is exponentially stable because

$$A + A^T = -cI. \tag{4.134}$$

The equality (4.134) guarantees that the Lyapunov function

$$V = \frac{1}{2}z^T z \tag{4.135}$$

has a negative definite time derivative

$$\dot{V} = -cz^T z. \tag{4.136}$$

Hence, the target system (4.129)–(4.131) is in a desirable form, but we still have to explain the relation between the change of variables (4.125)–(4.127) and the one used for PDEs in this chapter, as well as the relation between the structures in the plant (4.122)–(4.124) and the target system (4.129)–(4.131) relative to those of the PDE plants and the target systems in the chapter.

Let us first examine the change of variables $y \mapsto z$ in (4.125)–(4.127). This change of variables is clearly of the form $z = (I - K)[y]$, where I is the identity matrix and K is a "lower-triangular" nonlinear transformation. The lower-triangular structure of K in (4.125)–(4.127) is analogous to the Volterra structure of the spatially causal integral operator $\int_0^x k(x, y)u(y)\,dy$ in our change of variable $w(x) = u(x) - \int_0^x k(x, y)u(y)\,dy$ in this chapter. Another important feature of the change of variable (4.125)–(4.127) is that it is invertible, i.e., y can be expressed as a smooth function of z (to be specific, $y_1 = z_1$, $y_2 = z_2 - z_1^3 - cz_1$, and so on).

Next, let us examine the relation between the plant (4.122)–(4.124) and those studied in this chapter such as the reaction-diffusion system $u_t = u_{xx} + \lambda u$, as well as the relation between the target systems (4.129)–(4.131) and $w_t = w_{xx}$. The analogy between the target systems is particularly transparent because they both admit a simple 2-norm as a Lyapunov function, specifically,

$$\frac{d}{dt}\frac{1}{2}z^T z = -cz^T z \tag{4.137}$$

in the ODE case and

$$\frac{d}{dt}\frac{1}{2}\int_0^1 w(x)^2\,dx = -\int_0^1 w_x(x)^2\,dx \tag{4.138}$$

in the PDE case. A "finer" structural analogy, where one might expect the z-system to be a spatial discretization of the w-system, does not hold. If we discretize the PDE system $w_t = w_{xx}$, with boundary conditions $w(0,t) = w(1,t) = 0$, over a spatial grid with N points, we get the ODE system $\dot{w}_i = N^2(w_{i+1} - 2w_i + w_{i-1})$, which is different in structure from $\dot{z}_i = z_{i+1} - z_{i-1} - cz_i$, even after the N^2 factor (is absorbed into the time variable). This is where the subtle difference between ODE backstepping and PDE backstepping comes into play. The recursive procedure used for ODEs does not have a limit as the number of states goes to infinity. In contrast, the backstepping process for PDEs that we introduced in this chapter does have a limit, as we have proved in Section 4.4. Let us try to understand this difference by comparing the plant structure (4.122)–(4.124) with the plant structure $u_t = u_{xx} + \lambda u$. The former is dominated by a chain of integrators, while the latter is dominated by the diffusion operator. While the diffusion operator is a well-defined, meaningful object, an "infinite integrator chain" is not. It is for this reason that the infinite-dimensional backstepping design succeeds only if particular care is taken to convert the unstable parabolic PDE $u_t = u_{xx} + \lambda u$ into a stable target system $w_t = w_{xx}$ which is within the same PDE class, namely, parabolic. To put it in simpler words, we make sure to retain the ∂_{xx} term in the target system, even though it may be tempting to choose some other target system, such as, for example, the first-order hyperbolic (transport equation-like) PDE $w_t = w_x - cw$, which is more reminiscent of the ODE target system (4.129)–(4.131). If such an attempt is made, the derivation of the PDE conditions for the kernel $k(x, y)$ would not be successful and the matching of terms between the plant $u_t = u_{xx} + \lambda u$ and the target system $w_t = w_{xx} - cw$ would result in terms that cannot be cancelled.

Finally, let us explain the meaning of the term *backstepping*. In the ODE setting, this procedure is referred to as *integrator backstepping* because, as illustrated with the help of example (4.122)–(4.124), the design procedure propagates the feedback law synthesis "backwards" through a chain of integrators. Upon careful inspection of the change of variables (4.125)–(4.127), the first "step" of the backstepping procedure is to treat the state y_2 as the control input in the subsystem $\dot{y}_1 = y_2 + y_1^3$ and design the "control law" $y_2 = -y_1^3 - cy_1$, then "step back" through the integrator in the second subsystem $\dot{y}_2 = y_3 + y_2^3$ and design the "control" y_3 so that the error state $z_2 = y_2 - (-y_1^3 - cy_1)$ is forced to go zero, thus ensuring that the state y_2 acts (approximately) as the control $y_2 = -y_1^3 - cy_1$. This "backward stepping" through integrators continues until one encounters the actual control u in (4.128), which, in example (4.122)–(4.124), happens after two steps of backstepping. Even though in our *continuum* version of backstepping for PDEs there are no simple integrators to step through, the analogy with the method for ODEs is in the triangularity of the change of variable and the pursuit of a stable target system. For this reason, we retain the term *backstepping* for PDEs.

Backstepping for ODEs is applicable to a fairly broad class of ODE systems, which are referred to as *strict-feedback* systems. These systems are characterized by having a chain of integrators, the control appearing in the last equation, and having additional terms (linear or nonlinear) of a "lower-triangular" structure. In this lower-triangular structure the first equation depends only on the first state, the term in the second equation depends on the first and the second states, and so on. In the example (4.122)–(4.124) the cubic terms had a "diagonal" dependence on the states y_i, and thus their structure was lower-triangular, and hence the plant (4.122)–(4.124) was of strict-feedback type. The change of variables (4.125)–(4.127) has a general lower-triangular form. The capability of backstepping to deal with lower-triangular ODE structures has motivated our extension in Section 4.9 of PDE backstepping from reaction-diffusion systems (which are of a "diagonal" kind) to systems with lower-triangular strict-feedback terms $g(x)u(0,t)$ and $\int_0^x f(x,y)u(y,t)\,dy$. Such terms, besides being tractable in the backstepping method, happen to be essential in several applications, including flexible beams in Chapter 8 and Navier–Stokes equations in Chapter 11.

4.11 Notes and References

The material in this chapter is based on the ideas introduced in [113, 154, 156]. It is possible to modify the controllers developed in this chapter to achieve inverse optimality, which gives guaranteed robustness margins; see [154].

It is also possible to extend the approach to 2D and 3D PDEs in regular geometries; see [174] for a 2D extension in polar coordinates and Chapter 11 for a 3D infinite channel.

Some of the basic ingredients of backstepping have appeared in early works such as [38, 152], where integral transformations are used to solve PDEs and state controllability results, but not for design of feedback laws. It is interesting that these ideas appeared well before the development of finite-dimensional backstepping (though we became aware of them several years *after* the inception of our research program on backstepping control for PDEs, essentially rediscovering them, and arriving at the Volterra operator transformations from the finite-dimensional backstepping context). It is also curious that these powerful ideas were not pursued further after [38, 152].

Our very first attempt at developing continuum backstepping for PDEs was in [30], where we developed an explicit feedback law, backstepping transformation, and a Lyapunov function, but where the plant's level of open-loop instability was limited. Then, in [11, 26, 12] we turned our attention to a discretization-based (in space) approach, but the approach was dependent on the discretization scheme and did not yield convergent gain kernels when the discretization step $\delta x \to 0$, although, interestingly, the control input was nevertheless convergent, as it is an inner product of the gain kernel and the measured state of the system.

Exercises

4.1. For the plant

$$u_t = u_{xx} + \lambda u\,,$$
$$u_x(0) = 0\,,$$

design the Neumann stabilizing controller ($u_x(1)$ actuated).

Hint: Use the target system

$$\begin{aligned} w_t &= w_{xx}\,, \\ w_x(0) &= 0\,, \\ w_x(1) &= -\frac{1}{2}w(1)\,. \end{aligned} \tag{4.139}$$

This system is asymptotically stable (see Exercise 2.2). Note also that you do not need to find $k(x, y)$; it has already been found in Example 4.1. You need only use the condition (4.139) to derive the controller.

4.2. Find the PDE for the kernel $l(x, y)$ of the inverse transformation

$$u(x) = w(x) + \int_0^x l(x, y)w(y)\,dy\,,$$

which relates the systems u and w from Exercise 4.1. By comparison with the PDE for $k(x, y)$, show that

$$l(x, y) = -\lambda x \frac{J_1\left(\sqrt{\lambda(x^2 - y^2)}\right)}{\sqrt{\lambda(x^2 - y^2)}}\,.$$

4.3. Design the Dirichlet boundary controller for the heat equation

$$\begin{aligned} u_t &= u_{xx}\,, \\ u_x(0) &= -qu(0)\,. \end{aligned}$$

Follow these steps:

(1) Use the transformation

$$w(x) = u(x) - \int_0^x k(x, y)u(y)\,dy \tag{4.140}$$

to map the plant into the target system

$$w_t = w_{xx}\,, \tag{4.141}$$

$$w_x(0) = 0\,, \tag{4.142}$$

$$w(1) = 0\,. \tag{4.143}$$

Show that $k(x, y)$ satisfies the following PDE:

$$k_{xx}(x, y) = k_{yy}(x, y)\,, \tag{4.144}$$

$$k_y(x, 0) = -qk(x, 0)\,, \tag{4.145}$$

$$k(x, x) = -q\,. \tag{4.146}$$

(2) The general solution of the PDE (4.144) has the form $k(x, y) = \phi(x - y) + \psi(x + y)$, where ϕ and ψ are arbitrary functions. Using (4.146), it can be shown that $\psi \equiv 0$. Find ϕ from the conditions (4.145) and (4.146). Write the solution for $k(x, y)$.

(3) Write down the controller.

4.4. Show that the solution of the closed-loop system from Exercise 4.3 is

$$u(x,t) = 2\sum_{n=0}^{\infty} e^{-\sigma_n^2 t}\left(\sigma_n \cos(\sigma_n x) - q\sin(\sigma_n x)\right) \times \int_0^1 \frac{\sigma_n \cos(\sigma_n \xi) - q\sin(\sigma_n \xi) + (-1)^n q e^{q(1-\xi)}}{\sigma_n^2 + q^2} u_0(\xi)\, d\xi\,,$$

where $\sigma_n = \pi(2n+1)/2$. To do this, first write the solution of the system (4.141)–(4.143) (just set $\lambda = 0$ in the solution obtained in Exercise 3.1). Then use the transformation (4.140) with the $k(x, y)$ that you found in Exercise 4.3 to express the initial condition $w_0(x)$ in terms of $u_0(x)$ (you will need to change the order of integration in one of the terms to do this). Finally, write the solution for $u(x, t)$ using the inverse transformation

$$u(x) = w(x) - q\int_0^x w(y)\, dy$$

(i.e., $l(x, y) = -q$ in this problem; feel free to prove it).
Note that it is not possible to write a closed form solution for the open-loop plant, but it is possible to do so for the closed-loop system!

4.5. For the plant

$$u_t = u_{xx} + bu_x + \lambda u\,,$$
$$u_x(0) = -\frac{b}{2}u(0)\,,$$

design the Neumann stabilizing controller ($u_x(1)$ actuated).
Hint: By transforming the plant to a system without a b-term, reduce the problem to Exercise 4.1.

4.6. For the plant

$$u_t = u_{xx} + 3e^{2x}u(0)\,,$$
$$u_x(0) = 0\,,$$

design the Dirichlet stabilizing controller.

Chapter 5

Observer Design

The measurements in distributed parameter systems are rarely available across the domain. It is more common for the sensors to be placed only at the boundaries. This is particularly true in problems involving fluid flows (in applications such as aerodynamics, acoustics, chemical process control, etc.). Since state-feedback controllers developed thus far require the measurement of the state at each point in the domain, we need to design state observers.

5.1 Observer Design for PDEs

Consider the unstable heat equation with boundary actuation,

$$u_t = u_{xx} + \lambda u \,, \tag{5.1}$$

$$u_x(0) = 0 \,, \tag{5.2}$$

$$u(1) = U(t) \,, \tag{5.3}$$

where $U(t)$ may either represent open-loop forcing or be generated from a feedback law. Let us assume that only $u(0)$ is available for measurement. We will show that from this boundary information, it is possible to reconstruct the state in the domain.

We design the following observer for this plant:

$$\hat{u}_t = \hat{u}_{xx} + \lambda \hat{u} + p_1(x)[u(0) - \hat{u}(0)] \,, \tag{5.4}$$

$$\hat{u}_x(0) = p_{10}[u(0) - \hat{u}(0)] \,, \tag{5.5}$$

$$\hat{u}(1) = U(t). \tag{5.6}$$

Here the function $p_1(x)$ and the constant p_{10} are observer gains to be determined. It is helpful to note that the structure of the above observer mimics the well-known finite-dimensional observer format of "copy of the plant plus output injection." Indeed, for a finite-dimensional plant,

$$\dot{x} = Ax + Bu \,,$$
$$y = Cx \,,$$

the observer is

$$\dot{\hat{x}} = A\hat{x} + Bu + L(y - C\hat{x}), \tag{5.7}$$

where L is the observer gain and $L(y - C\hat{x})$ is the "output error injection." In (5.4) and (5.5) the observer gains $p_1(x)$ and p_{10} form an infinite-dimensional "vector" which corresponds to an analog of L.

Our objective is to find $p_1(x)$ and p_{10} such that $\hat{u}$ converges to u as time goes to infinity. To do this, we introduce the error variable

$$\tilde{u} = u - \hat{u} \tag{5.8}$$

and consider the error system

$$\tilde{u}_t = \tilde{u}_{xx} + \lambda\tilde{u} - p_1(x)\tilde{u}(0), \tag{5.9}$$

$$\tilde{u}_x(0) = -p_{10}\tilde{u}(0), \tag{5.10}$$

$$\tilde{u}(1) = 0. \tag{5.11}$$

With the (invertible) transformation

$$\tilde{u}(x) = \tilde{w}(x) - \int_0^x p(x,y)\tilde{w}(y)dy, \tag{5.12}$$

we transform the error system into the exponentially stable heat equation

$$\tilde{w}_t = \tilde{w}_{xx}, \tag{5.13}$$

$$\tilde{w}_x(0) = 0, \tag{5.14}$$

$$\tilde{w}(1) = 0. \tag{5.15}$$

Differentiating the transformation (5.12), we get

$$\begin{aligned}\tilde{u}_t(x) &= \tilde{w}_t(x) - \int_0^x p(x,y)\tilde{w}_{yy}(y)\,dy \\ &= \tilde{w}_t(x) - p(x,x)\tilde{w}_x(x) + p(x,0)\tilde{w}_x(0) + p_y(x,x)\tilde{w}(x) \\ &\quad - p_y(x,0)\tilde{w}(0) - \int_0^x p_{yy}(x,y)\tilde{w}(y)\,dy,\end{aligned} \tag{5.16}$$

$$\begin{aligned}\tilde{u}_{xx}(x) &= \tilde{w}_{xx}(x) - \tilde{w}(x)\frac{d}{dx}p(x,x) - p(x,x)\tilde{w}_x(x) \\ &\quad - p_x(x,x)\tilde{w}(x) - \int_0^x p_{xx}(x,y)\tilde{w}(y)\,dy.\end{aligned} \tag{5.17}$$

Subtracting (5.17) from (5.16), we obtain

$$\begin{aligned}&\lambda\left(\tilde{w}(x) - \int_0^x p(x,y)\tilde{w}(y)\,dy\right) - p_1(x)\tilde{w}(0) \\ &= 2\tilde{w}(x)\frac{d}{dx}p(x,x) - p_y(x,0)\tilde{w}(0) + \int_0^x (p_{xx}(x,y) - p_{yy}(x,y))\tilde{w}(y)\,dy.\end{aligned} \tag{5.18}$$

For the last equality to hold, the following three conditions must be satisfied:

$$p_{xx}(x,y) - p_{yy}(x,y) = -\lambda p(x,y)\,, \tag{5.19}$$

$$\frac{d}{dx}p(x,x) = \frac{\lambda}{2}\,, \tag{5.20}$$

$$p_1(x) = p_y(x,0)\,. \tag{5.21}$$

The boundary conditions (5.10) and (5.11) provide two more conditions as follows:

$$p_{10} = p(0,0)\,, \tag{5.22}$$

$$p(1,y) = 0\,. \tag{5.23}$$

The condition (5.22) is obtained by differentiating (5.12) with respect to x, setting $x = 0$, and substituting (5.10) and (5.14) in the resulting equation. The condition (5.23) is obtained by setting $x = 1$ in (5.12) and substituting (5.11) and (5.15) in the resulting equation.

Let us solve (5.20) and (5.23) for $p(x,x)$ and combine the result with (5.19) and (5.23) as follows:

$$p_{xx}(x,y) - p_{yy}(x,y) = -\lambda p(x,y)\,, \tag{5.24}$$

$$p(1,y) = 0\,, \tag{5.25}$$

$$p(x,x) = \frac{\lambda}{2}(x-1)\,. \tag{5.26}$$

These three conditions form a well posed PDE which we can solve explicitly. To do this, we make a change of variables

$$\bar{x} = 1 - y, \quad \bar{y} = 1 - x, \quad \bar{p}(\bar{x},\bar{y}) = p(x,y)\,, \tag{5.27}$$

which gives the following PDE:

$$\bar{p}_{\bar{x}\bar{x}}(\bar{x},\bar{y}) - \bar{p}_{\bar{y}\bar{y}}(\bar{x},\bar{y}) = \lambda p(\bar{x},\bar{y})\,, \tag{5.28}$$

$$\bar{p}(\bar{x},0) = 0\,, \tag{5.29}$$

$$\bar{p}(\bar{x},\bar{x}) = -\frac{\lambda}{2}x\,. \tag{5.30}$$

This PDE was solved in Chapter 4, and its solution is

$$\bar{p}(\bar{x},\bar{y}) = -\lambda\bar{y}\frac{I_1(\sqrt{\lambda(\bar{x}^2-\bar{y}^2)})}{\sqrt{\lambda(\bar{x}^2-\bar{y}^2)}} \tag{5.31}$$

or, in the original variables,

$$p(x,y) = -\lambda(1-x)\frac{I_1(\sqrt{\lambda(2-x-y)(x-y)})}{\sqrt{\lambda(2-x-y)(x-y)}}\,. \tag{5.32}$$

The observer gains, obtained using (5.21) and (5.22) are

$$p_1(x) = p_y(x,0) = \frac{\lambda(1-x)}{x(2-x)}I_2\left(\sqrt{\lambda x(2-x)}\right), \tag{5.33}$$

$$p_{10} = p(0,0) = -\frac{\lambda}{2}\,. \tag{5.34}$$

5.2 Output Feedback

The exponentially convergent observer developed in the last section is independent of the control input and can be used with any controller. In this section we combine it with the backstepping controller developed in Chapter 4 to solve the output-feedback problem.

For linear systems, the separation principle (or "certainty equivalence") holds; i.e., the combination of a separately designed state feedback controller and observer results in a stabilizing output-feedback controller. Next, we establish the separation principle for our observer-based output-feedback design.

It is straightforward to show that the observer and control backstepping transformations (5.12) and

$$\hat{w}(x) = \hat{u}(x) - \int_0^x k(x,y)\hat{u}(y)\,dy\,, \tag{5.35}$$

$$\hat{u}(x) = \hat{w}(x) + \int_0^x l(x,y)\hat{w}(y)\,dy \tag{5.36}$$

map the closed-loop system consisting of the observer error PDE and the observer into the following target system:

$$\begin{aligned}
\hat{w}_t &= \hat{w}_{xx} + \left\{p_1(x) - \int_0^x k(x,y)p_1(y)\,dy + p_{10}k(x,0)\right\}\tilde{w}(0)\,, && (5.37)\\
\hat{w}_x(0) &= p_{10}\tilde{w}(0)\,, && (5.38)\\
\hat{w}(1) &= 0\,, && (5.39)\\
\tilde{w}_t &= \tilde{w}_{xx}\,, && (5.40)\\
\tilde{w}_x(0) &= 0\,, && (5.41)\\
\tilde{w}(1) &= 0\,, && (5.42)
\end{aligned}$$

where $k(x,y)$ is the kernel of the control transformation and $p_1(x)$, p_{10} are observer gains. The $\tilde{w}$-system and the homogeneous part of the $\hat{w}$-system (without $\tilde{w}(0,t)$) are exponentially stable heat equations. To show that the system $(\hat{w}, \tilde{w})$ is exponentially stable, we use the weighted Lyapunov function

$$V = \frac{A}{2}\int_0^1 \tilde{w}(x)^2\,dx + \frac{1}{2}\int_0^1 \hat{w}(x)^2\,dx\,, \tag{5.43}$$

where A is the weighting constant to be chosen later. Taking the time derivative of (5.43), we get

$$\begin{aligned}
\dot{V} = {}& -A\int_0^1 \tilde{w}_x(x)^2\,dx - p_{10}\hat{w}(0)\tilde{w}(0) - \int_0^1 \hat{w}_x(x)^2\,dx\\
& + \tilde{w}(0)\int_0^1 \hat{w}(x)\left\{p_1(x) - \int_0^x k(x,y)p_1(y)\,dy + p_{10}k(x,0)\right\}dx\,.
\end{aligned}$$

Using the Poincaré and Young inequalities, we estimate

$$-p_{10}\hat{w}(0)\tilde{w}(0) \le \frac{1}{4}\hat{w}(0)^2 + p_{10}^2\tilde{w}(0)^2 \le \frac{1}{4}\int_0^1 \hat{w}_x(x)^2\,dx + p_{10}^2\int_0^1 \tilde{w}_x(x)^2\,dx$$

and

$$\tilde{w}(0)\int_0^1 \hat{w}(x)\left\{p_1(x)-\int_0^x k(x,y)p_1(y)\,dy+p_{10}k(x,0)\right\}dx$$
$$\le \frac{1}{4}\int_0^1 \hat{w}_x(x)^2\,dx + B^2\int_0^1 \tilde{w}_x(x)^2\,dx\,,$$

where $B = \max_{x\in[0,1]}\{p_1(x)-\int_0^x k(x,y)p_1(y)\,dy+p_{10}k(x,0)\}$. With these estimates, we obtain

$$\dot{V} \le -(A-B^2-p_{10}^2)\int_0^1 \tilde{w}_x(x)^2\,dx - \frac{1}{2}\int_0^1 \hat{w}_x(x)^2\,dx$$
$$\le -\frac{1}{4}(A-B^2-p_{10}^2)\int_0^1 \tilde{w}(x)^2\,dx - \frac{1}{8}\int_0^1 \hat{w}(x)^2\,dx\,.$$

Taking $A = 2(B^2+p_{10}^2)$, we get

$$\dot{V} \le -\frac{1}{4}V\,.$$

Hence, the system $(\hat{w},\,\tilde{w})$ is exponentially stable. The system $(\hat{u},\,\tilde{u})$ is also exponentially stable since it is related to $(\hat{w},\,\tilde{w})$ by the invertible coordinate transformations (5.12) and (5.36). We have thus proved that the closed-loop system consisting of the plant with backstepping controller and observer is exponentially stable.

In Table 5.1 we summarize the output-feedback design for the case when the sensor and actuator are placed at opposite boundaries ("anticollocated" setup).

5.3 Observer Design for Collocated Sensor and Actuator

When the sensor and actuator are placed at the same boundary, the observer design is slightly different. Consider the same plant as in the previous section, but with the following collocated measurement and actuation:

$$u_t = u_{xx} + \lambda u\,, \tag{5.50}$$
$$u_x(0) = 0\,, \tag{5.51}$$
$$u(1) = U(t)\,, \tag{5.52}$$
$$u_x(1) \;-\; \text{measurement.}$$

We design the following observer:

$$\hat{u}_t = \hat{u}_{xx} + \lambda\hat{u} + p_1(x)[u_x(1)-\hat{u}_x(1)]\,, \tag{5.53}$$
$$\hat{u}_x(0) = 0\,, \tag{5.54}$$
$$\hat{u}(1) = U(t) + p_{10}[u_x(1)-\hat{u}_x(1)]\,. \tag{5.55}$$

Note here that the output injection is placed at the same boundary at which the sensor is located, i.e., at $x=1$ (not at $x=0$ as in the anticollocated case).

Table 5.1. *Output-feedback design for anticollocated setup.*

Plant:

$$u_t = u_{xx} + \lambda u \,, \tag{5.44}$$
$$u_x(0) = 0 \,. \tag{5.45}$$

Observer:

$$\hat{u}_t = \hat{u}_{xx} + \lambda \hat{u} + \frac{\lambda(1-x)}{x(2-x)} I_2\left(\sqrt{\lambda x(2-x)}\right)[u(0) - \hat{u}(0)] \,, \tag{5.46}$$
$$\hat{u}_x(0) = -\frac{\lambda}{2}[u(0) - \hat{u}(0)] \,, \tag{5.47}$$
$$\hat{u}(1) = -\int_0^1 \lambda \frac{I_1(\sqrt{\lambda(1-y^2)})}{\sqrt{\lambda(1-y^2)}} \hat{u}(y)\, dy \,. \tag{5.48}$$

Controller:

$$u(1) = -\int_0^1 \lambda \frac{I_1(\sqrt{\lambda(1-y^2)})}{\sqrt{\lambda(1-y^2)}} \hat{u}(y)\, dy \,. \tag{5.49}$$

Introducing the error $\tilde{u} = u - \hat{u}$, we get the error system

$$\tilde{u}_t = \tilde{u}_{xx} + \lambda \tilde{u} - p_1(x)\tilde{u}_x(1) \,, \tag{5.56}$$
$$\tilde{u}_x(0) = 0 \,, \tag{5.57}$$
$$\tilde{u}(1) = -p_{10}\tilde{u}_x(1) \,. \tag{5.58}$$

We use the transformation

$$\tilde{u}(x) = \tilde{w}(x) - \int_x^1 p(x, y)\tilde{w}(y)dy \tag{5.59}$$

to convert the error system into the following exponentially stable target system:

$$\tilde{w}_t = \tilde{w}_{xx} \,, \tag{5.60}$$
$$\tilde{w}_x(0) = 0 \,, \tag{5.61}$$
$$\tilde{w}(1) = 0. \tag{5.62}$$

Note that the integral in the transformation has limits from x to 1 instead of the usual limits from 0 to x.

By substituting (5.59) into the plant, we get the following set of conditions on the observer kernel $p(x, y)$ in the form of a hyperbolic PDE:

$$p_{xx}(x, y) - p_{yy}(x, y) = -\lambda p(x, y) \tag{5.63}$$

with the boundary conditions

$$p_x(0, y) = 0\,, \tag{5.64}$$

$$p(x, x) = -\frac{\lambda}{2}x \tag{5.65}$$

that yield

$$\tilde{w}_t = \tilde{w}_{xx} + [p(x, 1) - p_1(x)]\tilde{w}_x(1)\,, \tag{5.66}$$

$$\tilde{w}_x(0) = 0\,, \tag{5.67}$$

$$\tilde{w}(1) = -p_{10}w_x(1)\,. \tag{5.68}$$

From the comparison of this system with (5.60)–(5.62), it follows that the observer gains should be chosen as

$$p_1(x) = p(x, 1), \qquad p_{10} = 0\,. \tag{5.69}$$

To solve the PDE (5.63)–(5.65) we introduce the change of variables

$$\bar{x} = y, \quad \bar{y} = x, \quad \bar{p}(\bar{x}, \bar{y}) = p(x, y)$$

to get

$$\bar{p}_{\bar{x}\bar{x}}(\bar{x}, \bar{y}) - \bar{p}_{\bar{y}\bar{y}}(\bar{x}, \bar{y}) = \lambda \bar{p}(\bar{x}, \bar{y}) \tag{5.70}$$

$$\bar{p}_{\bar{y}}(\bar{x}, 0) = 0\,, \tag{5.71}$$

$$\bar{p}(\bar{x}, \bar{x}) = -\frac{\lambda}{2}\bar{x}\,. \tag{5.72}$$

This PDE was solved in Chapter 4 and its solution is

$$\begin{aligned} p(\bar{x}, \bar{y}) &= -\lambda \bar{x}\frac{I_1(\sqrt{\lambda(\bar{x}^2 - \bar{y}^2)})}{\sqrt{\lambda(\bar{x}^2 - \bar{y}^2)}} \\ &= -\lambda y\frac{I_1(\sqrt{\lambda(y^2 - x^2)})}{\sqrt{\lambda(y^2 - x^2)}}\,. \end{aligned}$$

Therefore, the observer gains are

$$p_1(x) = -\lambda\frac{I_1(\sqrt{\lambda(1 - x^2)})}{\sqrt{\lambda(1 - x^2)}}\,, \tag{5.73}$$

and $p_{10} = 0$.

Remark 5.1. The fact that $p_1(x) = k(1, x)$ demonstrates the duality between observer and control designs—a well-known concept in linear control theory for finite-dimensional systems.

Remark 5.2. We assumed in this chapter that the decay rates of the observer and controller are the same. One can easily modify the designs to achieve the typically desired time scale decomposition, where the observer decays faster than the controller.

Table 5.2. *Output-feedback design for collocated setup.*

Plant:

$$u_t = u_{xx} + \lambda u \,, \tag{5.74}$$
$$u_x(0) = 0 \,. \tag{5.75}$$

Observer:

$$\hat{u}_t = \hat{u}_{xx} + \lambda \hat{u} - \lambda \frac{I_1(\sqrt{\lambda(1-x^2)})}{\sqrt{\lambda(1-x^2)}}[u_x(1) - \hat{u}_x(1)] \,, \tag{5.76}$$
$$\hat{u}_x(0) = 0 \,, \tag{5.77}$$
$$\hat{u}(1) = -\int_0^1 \lambda \frac{I_1(\sqrt{\lambda(1-y^2)})}{\sqrt{\lambda(1-y^2)}} \hat{u}(y)\, dy \,. \tag{5.78}$$

Controller:

$$u(1) = -\int_0^1 \lambda \frac{I_1(\sqrt{\lambda(1-y^2)})}{\sqrt{\lambda(1-y^2)}} \hat{u}(y)\, dy \,. \tag{5.79}$$

As in the previous section, the collocated observer and controller are combined into an output-feedback compensator. The summary of the output-feedback design is presented in Table 5.2.

5.4 Compensator Transfer Function

When both the controller and the observer are given explicitly, one can derive a frequency domain representation of the compensator.

To illustrate this, we consider the following PDE:

$$u_t = u_{xx} + g u(0) \,, \tag{5.80}$$
$$u_x(0) = 0 \tag{5.81}$$

with $u(0)$ measured and $u(1)$ actuated. We first derive the transfer function of the open-loop plant. Taking the Laplace transform of (5.80) and (5.81), we get

$$s u(x,s) = u''(x,s) + g u(0,s) \,, \tag{5.82}$$
$$u'(0,s) = 0 \,. \tag{5.83}$$

The general solution for this second-order ODE is given by

$$u(x,s) = A \sinh(\sqrt{s}x) + B \cosh(\sqrt{s}x) + \frac{g}{s} u(0,s) \,, \tag{5.84}$$

where A and B are constants to be determined. From the boundary condition (5.83) we have

$$u'(0,s) = A\sqrt{s} = 0 \Rightarrow A = 0\,. \tag{5.85}$$

By setting $x = 0$ in (5.84), we find B:

$$B = u(0,s)\left(1 - \frac{g}{s}\right). \tag{5.86}$$

Hence, we get

$$u(x,s) = u(0,s)\left[\frac{g}{s} + \left(1 - \frac{g}{s}\right)\cosh(\sqrt{s}x)\right].$$

Setting $x = 1$, we obtain

$$\boxed{u(0,s) = \frac{s}{g + (s-g)\cosh(\sqrt{s})}u(1,s)\,.} \tag{5.87}$$

This plant has no zeros and has an infinite relative degree. Using a Taylor expansion of the cosh term, we get an approximate expression for the plant transfer function,

$$\frac{u(0,s)}{u(1,s)} \approx \frac{1}{1 - \frac{g}{2} + \left(\frac{1}{2} - \frac{g}{4!}\right)s + \left(\frac{1}{4!} - \frac{g}{6!}\right)s^2 + \cdots}\,. \tag{5.88}$$

Let us now derive the frequency domain representation of the compensator. The observer PDE is given by

$$\hat{u}_t = \hat{u}_{xx} + gu(0)\,, \tag{5.89}$$

$$\hat{u}_x(0) = 0\,, \tag{5.90}$$

$$\hat{u}(1) = -\int_0^1 \sqrt{g}\sinh(\sqrt{g}(1-y))\hat{u}(y)dy\,. \tag{5.91}$$

Applying the Laplace transform, we get

$$s\hat{u}(x,s) = \hat{u}''(x,s) + gu(0,s)\,, \tag{5.92}$$

$$\hat{u}'(0,s) = 0\,, \tag{5.93}$$

$$\hat{u}(1,s) = -\int_0^1 \sqrt{g}\sinh(\sqrt{g}(1-y))\hat{u}(y,s)dy\,. \tag{5.94}$$

It is easy to show that the general solution of PDE (5.92) with boundary condition (5.93) is given by

$$\hat{u}(x,s) = \hat{u}(0,s)\cosh(\sqrt{s}x) + \frac{g}{s}\left(1 - \cosh(\sqrt{s}x)\right)u(0,s)\,. \tag{5.95}$$

Substituting (5.95) into (5.94), and evaluating the integral, we express $\hat{u}(0,s)$ as a function of $u(0,s)$:

$$\hat{u}(0,s) = \frac{\cosh(\sqrt{s}) - \cosh(\sqrt{g})}{s\cosh(\sqrt{s}) - g\cosh(\sqrt{g})}gu(0,s)\,. \tag{5.96}$$

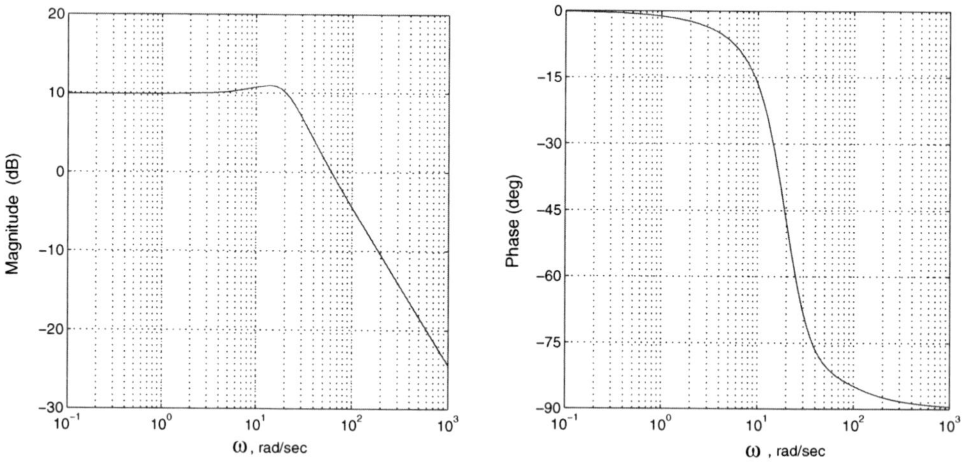

Figure 5.1. *Bode plots of the compensator* (5.97) .

Setting $x = 1$ in (5.95) and using (5.96), we get the desired transfer function of the compensator:

$$u(1,s) = \frac{g}{s}\left(-1 + \frac{(s-g)\cosh(\sqrt{s})\cosh(\sqrt{g})}{s\cosh(\sqrt{s}) - g\cosh(\sqrt{g})}\right)u(0,s)\,. \tag{5.97}$$

The Bode plots of the compensator for $g = 8$ are shown in Figure 5.1. One can see that it can be approximated by a second order, relative degree one transfer function. For example, a pretty good approximation would be

$$C(s) \approx 60\frac{s+17}{s^2+25s+320}\,. \tag{5.98}$$

One might find the observation that the unstable, infinite relative degree plant (5.80), (5.81) is stabilizable using a relative degree one compensator to be intriguing, perhaps even fascinating, as one could not expect this to be possible in general. In fact, one also should not generally expect such a simple compensator to arise from the backstepping design. Even for this particular simple plant, which can have only one unstable eigenvalue, a higher relative degree compensator might very well be needed for a higher value of g in the plant.

A legitimate question to raise is that of a possible "spill over" effect of the higher-order modes under the truncated compensator (5.98). Such a phenomenon, however, is atypical for parabolic PDEs, in which all but possibly very few eigenvalues are negative real and very large in absolute value. In this situation, the singular perturbation principle (where the boundary layer model is exponentially stable) allows us to achieve stabilization with a compensator of finite order. The actual proof of this, however, is not elementary and, as such, is beyond the scope of our presentation.

5.5 Notes and References

The material in this chapter is based on ideas introduced in [155].

Exercises

5.1. Design an observer for the following system:

$$\begin{aligned} u_t &= u_{xx}\,, \\ u_x(0) &= -qu(0)\,, \\ u(1) &= U(t) \end{aligned}$$

with only $u_x(1)$ available for measurement.
Follow these steps:

(1) Write down the observer for this system, with output injection entering the PDE and the boundary condition at $x = 1$.

(2) Use the transformation

$$\tilde{u}(x) = \tilde{w}(x) - \int_x^1 p(x, y)\tilde{w}(y)\,dy \tag{5.99}$$

to map the error system into the target system

$$\tilde{w}_t = \tilde{w}_{xx}\,, \tag{5.100}$$

$$\tilde{w}_x(0) = 0\,, \tag{5.101}$$

$$\tilde{w}(1) = 0\,. \tag{5.102}$$

Show that $p(x, y)$ satisfies the PDE

$$p_{xx}(x, y) = p_{yy}(x, y)\,, \tag{5.103}$$

$$p_x(0, y) = -qp(0, y)\,, \tag{5.104}$$

$$p(x, x) = -q \tag{5.105}$$

and that the observer gains are given by $p_{10} = 0$ and $p_1(x) = p(x, 1)$.

(3) Solve the PDE (5.103)–(5.105) for $p(x, y)$ (look for the solution in the form $p(x, y) = \phi(y - x)$). Find $p_1(x)$.

5.2. Find the frequency domain representation of the plant

$$\begin{aligned} u_t &= u_{xx}\,, \\ u_x(0) &= -qu(0)\,, \\ u(1) &= U(t) \end{aligned}$$

with $u(0)$ measured and $u(1)$ actuated; i.e., find $G(s)$ such that $u(0, s) = G(s)U(s)$.

Chapter 6

Complex-Valued PDEs: Schrödinger and Ginzburg–Landau Equations

In this chapter we extend the designs developed in Chapter 4 to the case of "parabolic-like" plants with a complex-valued state. Such plants can also be viewed as two coupled PDEs. We consider two classes of such plants: the Ginzburg–Landau system (Section 6.2) and its special case, the Schrödinger equation (Section 6.1). For breadth of illustration, the Schrödinger equation is treated as a single complex-valued equation, whereas the Ginzburg–Landau equation is treated as two coupled PDEs.

In what follows, j denotes the imaginary unit, $\sqrt{-1}$.

6.1 Schrödinger Equation

The simplest complex-valued PDE is a linearized Schrödinger equation

$$v_t = -j v_{xx}\,, \tag{6.1}$$

$$v_x(0) = 0\,, \tag{6.2}$$

where $v(1)$ is actuated and $v(x,t)$ is a complex-valued function. Without control, this system displays oscillatory behavior and is not asymptotically stable.

Interestingly, the Schrödinger equation (6.1) is equivalent to the Euler–Bernoulli beam equation (8.1), considered later in Chapter 8. (Both the real part of v and the imaginary part of v satisfy separate Euler–Bernoulli beam equations.)

A striking fact is that the stabilization problem for (6.1) and (6.2) is easily solved using our control design for parabolic PDEs in Chapter 4. Let us formally think of $-j$ as the diffusion coefficient in the reaction-advection-diffusion equation (4.77) (with the advection and reaction coefficients being zero) and follow the control design developed in Section 4.7. We use the transformation

$$w(x) = v(x) - \int_0^x k(x,y)v(y)\,dy\,, \tag{6.3}$$

along with boundary feedback

$$v(1) = \int_0^1 k(1,y)v(y)\,dy\,, \tag{6.4}$$

where now $k(x, y)$ is a complex-valued control gain for mapping (6.1) and (6.2) into the target system (4.86) with $\varepsilon = -j$ and with the modified boundary condition at $x = 0$:

$$w_t = -jw_{xx} - cw\,, \tag{6.5}$$

$$w_x(0) = 0\,, \tag{6.6}$$

$$w(1) = 0\,, \tag{6.7}$$

which is exponentially stable for $c > 0$ (see Exercise 6.1).

The gain kernel PDE takes the form

$$k_{xx}(x, y) - k_{yy}(x, y) = cjk(x, y)\,, \tag{6.8}$$

$$k_y(x, 0) = 0\,, \tag{6.9}$$

$$k(x, x) = -\frac{cj}{2}x\,. \tag{6.10}$$

The solution to this PDE was obtained in Example 4.1 (set $\lambda = cj$). The control gain is

$$k(x, y) = -cjx\frac{I_1\left(\sqrt{cj(x^2 - y^2)}\right)}{\sqrt{cj(x^2 - y^2)}} \tag{6.11}$$

$$= x\sqrt{\frac{c}{2\left(x^2 - y^2\right)}}\left[(j-1)\mathrm{ber}_1\left(\sqrt{c\left(x^2 - y^2\right)}\right) - (1+j)\mathrm{bei}_1\left(\sqrt{c\left(x^2 - y^2\right)}\right)\right]$$

where $\mathrm{ber}_1(\cdot)$ and $\mathrm{bei}_1(\cdot)$ are the Kelvin functions, which are defined as

$$\mathrm{ber}_1(x) = -\mathrm{Im}\left\{I_1\left(\frac{1+j}{\sqrt{2}}x\right)\right\}, \tag{6.12}$$

$$\mathrm{bei}_1(x) = \mathrm{Re}\left\{I_1\left(\frac{1+j}{\sqrt{2}}x\right)\right\}. \tag{6.13}$$

In Figure 6.1 the open-loop and closed-loop behavior of the Schrödinger equation are shown (only for the real part of the state, the imaginary part is similar). The control effort is shown in Figure 6.2.

Remark 6.1. A simpler boundary feedback, $u_x(1, t) = -jc_1u(1, t)$, $c_1 > 0$, for the Schrödinger equation (6.1), (6.2) also achieves exponential stabilization. However, it cannot provide an arbitrary decay rate of the closed-loop system. Under this feedback, increasing c_1 moves the eigenvalues to the left in the complex plane up to some point, and then the first several eigenvalues return to the imaginary axis. In contrast, the backstepping design moves all the eigenvalues arbitrarily to the left. It is in fact possible to combine both designs by replacing the boundary condition (6.7) of the target system (6.5)–(6.7) with

$$w_x(1) = -jc_1w(1)\,. \tag{6.14}$$

Then the combined controller for the Schrödinger equation becomes

$$v_x(1) = -j\left(\frac{c}{2} + c_1\right)v(1) + \int_0^1 (k_x(1, y) + jc_1k(1, y))v(y)\,dy\,. \tag{6.15}$$

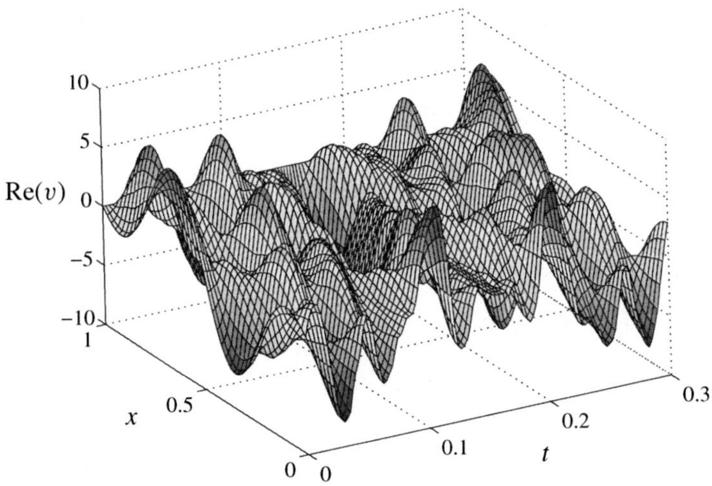

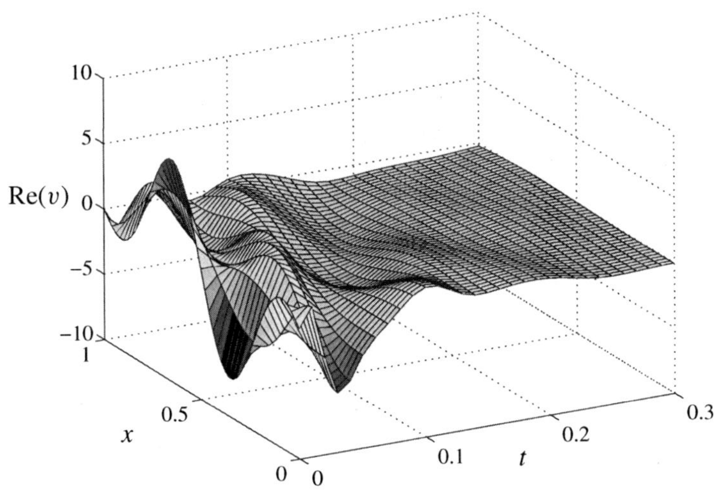

Figure 6.1. *The open-loop (top) and closed-loop (bottom) responses of the Schrödinger equation.*

A more general version of the Schrödinger equation is the Ginzburg–Landau equation, which is considered next.

6.2 Ginzburg–Landau Equation

In flows past submerged obstacles, the phenomenon of vortex shedding occurs when the Reynolds number is sufficiently large. A prototype model flow for studying vortex shedding is the flow past a two-dimensional circular cylinder (Figure 6.3). The vortices, which are alternately shed from the upper and lower sides of the cylinder, induce an undesirable

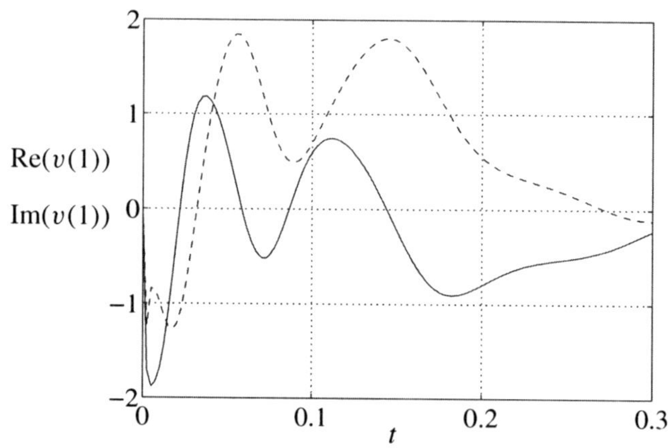

Figure 6.2. *Real (solid) and imaginary (dashed) parts of the control effort for the Schrödinger equation.*

periodic force that acts on the cylinder. The dynamics of the cylinder wake, often referred to as the von Karman vortex street, are governed by the Navier–Stokes equations. However, a simpler model exists in the form of the Ginzburg–Landau equation,

$$\frac{\partial A}{\partial t} = a_1 \frac{\partial^2 A}{\partial \breve{x}^2} + a_2(\breve{x}) \frac{\partial A}{\partial \breve{x}} + a_3(\breve{x}) A \tag{6.16}$$

for $\breve{x} \in (x_d, 1)$, with boundary conditions

$$A(x_d, t) = 0, \tag{6.17}$$

$$A(1, t) = u(t), \tag{6.18}$$

where A is a complex-valued function of $(\breve{x}, t)$, which is related to the transverse fluctuating velocity, $\mathrm{Re}(a_1) > 0$, and u is the control input. The coefficients a_1, a_2, a_3 are fitted to data from laboratory experiments. Note that the fluid flows in the negative $\breve{x}$ direction to fit with our notation in the other chapters of this book, where the control input is at the right boundary.

The model (6.16)–(6.18) is in fact a simplified version of the full nonlinear Ginzburg–Landau model. First, the nonlinear model is linearized around the zero solution, and then the (originally infinite) domain is truncated on both sides. The upstream subsystem (beyond $\breve{x} = 1$) is discarded since it is an approximately uniform flow, and the downstream subsystem is truncated at some $x_d \in (-\infty, 1)$, which can be selected to achieve any desired level of accuracy (because sufficiently far downstream from the cylinder, the flow again becomes uniform).

Our objective is to prevent vortex shedding by applying feedback control at $\breve{x} = 1$ using microjet actuators distributed on the cylinder surface. The measurements are assumed to be taken at the location of the cylinder only (via pressure sensors), collocated with actuation. Although the "anticollocated" case (with measurements taken at $\breve{x} = x_d$) can also be solved by our method, we focus on the collocated case since it avoids the use of unrealistic mid-flow measurements.

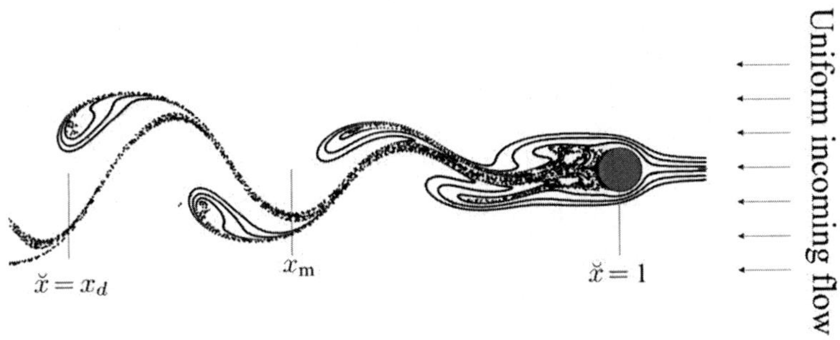

Figure 6.3. *Vortex shedding in the 2D flow past a cylinder.*

6.2.1 Problem Formulation as Two Coupled PDEs

We now rewrite equation (6.16) in the form of two coupled PDEs with real states and coefficients and transform the domain $[x_d, 1]$ into the domain $[0, 1]$. Define

$$\rho(x,t) = \mathrm{Re}(B(x,t)) = \left(B(x,t) + \bar{B}(x,t)\right)/2 \tag{6.19}$$

$$\iota(x,t) = \mathrm{Im}(B(x,t)) = \left(B(x,t) - \bar{B}(x,t)\right)/(2j)\,, \tag{6.20}$$

where

$$x = \frac{\breve{x} - x_d}{1 - x_d}\,, \tag{6.21}$$

$$B(x,t) = A(\breve{x},t)\exp\left(\frac{1}{2a_1}\int_{x_d}^{\breve{x}} a_2(\tau)\,d\tau\right)\,, \tag{6.22}$$

and $\bar{}$ denotes complex conjugation. Equation (6.16) becomes

$$\rho_t = a_R\rho_{xx} + b_R(x)\rho - a_I\iota_{xx} - b_I(x)\iota\,, \tag{6.23}$$

$$\iota_t = a_I\rho_{xx} + b_I(x)\rho + a_R\iota_{xx} + b_R(x)\iota \tag{6.24}$$

for $x \in (0, 1)$, with boundary conditions

$$\rho(0,t) = 0\,, \tag{6.25}$$

$$\iota(0,t) = 0\,, \tag{6.26}$$

$$\rho(1,t) = u_R(t)\,, \tag{6.27}$$

$$\iota(1,t) = u_I(t)\,, \tag{6.28}$$

where

$$a_R = \frac{\mathrm{Re}(a_1)}{(1-x_d)^2}\,, \qquad a_I = \frac{\mathrm{Im}(a_1)}{(1-x_d)^2}\,, \tag{6.29}$$

and

$$b_R(x) = \mathrm{Re}\left(a_3(\breve{x}) - \frac{1}{2}a_2'(\breve{x}) - \frac{1}{4a_1}a_2^2(\breve{x})\right)\,, \tag{6.30}$$

$$b_I(x) = \mathrm{Im}\left(a_3(\breve{x}) - \frac{1}{2}a_2'(\breve{x}) - \frac{1}{4a_1}a_2^2(\breve{x})\right)\,. \tag{6.31}$$

6.2.2 Control Design

First, we design the controllers $u_R(t)$ and $u_I(t)$ to stabilize the system (6.23)–(6.24), assuming full-state measurements.

Since the plant consists of two coupled PDEs, we introduce two backstepping transformations:

$$\breve{\rho}(x,t) = \rho(x,t) - \int_0^x \left[k(x,y)\,\rho(y,t) + k_c(x,y)\,\iota(y,t)\right]dy, \tag{6.32}$$

$$\breve{\iota}(x,t) = \iota(x,t) - \int_0^x \left[-k_c(x,y)\,\rho(y,t) + k(x,y)\,\iota(y,t)\right]dy \tag{6.33}$$

with two gain kernels $k(x,y)$ and $k_c(x,y)$.

For the target system we choose (6.23)–(6.25), with $b_R(x)$ and $b_I(x)$ replaced by the different functions $f_R(x)$ and $f_I(x)$:

$$\breve{\rho}_t = a_R\breve{\rho}_{xx} + f_R(x)\,\breve{\rho} - a_I\breve{\iota}_{xx} - f_I(x)\,\breve{\iota}, \tag{6.34}$$

$$\breve{\iota}_t = a_I\breve{\rho}_{xx} + f_I(x)\,\breve{\rho} + a_R\breve{\iota}_{xx} + f_R(x)\,\breve{\iota} \tag{6.35}$$

for $x \in (0,1)$, with boundary conditions

$$\breve{\rho}(0,t) = \breve{\iota}(0,t) = 0, \tag{6.36}$$

$$\breve{\rho}(1,t) = \breve{\iota}(1,t) = 0. \tag{6.37}$$

The controller is obtained by setting $x = 1$ in (6.32) and (6.33):

$$u_R(t) = \int_0^1 \left[k(1,y)\rho(y,t) + k_c(1,y)\iota(y,t)\right]dy, \tag{6.38}$$

$$u_I(t) = \int_0^1 \left[-k_c(1,y)\rho(y,t) + k(1,y)\iota(y,t)\right]dy. \tag{6.39}$$

One can show that the pair of kernels, $k(x,y)$ and $k_c(x,y)$, satisfy the following two coupled PDEs:

$$k_{xx} = k_{yy} + \beta(x,y)k + \beta_c(x,y)k_c, \tag{6.40}$$

$$k_{c,xx} = k_{c,yy} - \beta_c(x,y)k + \beta(x,y)k_c \tag{6.41}$$

for $0 < y < x < 1$, with boundary conditions

$$k(x,0) = 0, \tag{6.42}$$

$$k_c(x,0) = 0, \tag{6.43}$$

$$k(x,x) = -\frac{1}{2}\int_0^x \beta(\gamma,\gamma)d\gamma, \tag{6.44}$$

$$k_c(x,x) = \frac{1}{2}\int_0^x \beta_c(\gamma,\gamma)d\gamma, \tag{6.45}$$

where

$$\beta(x,y) = \frac{a_R\,(b_R(y) - f_R(x)) + a_I\,(b_I(y) - f_I(x))}{a_R^2 + a_I^2}, \tag{6.46}$$

$$\beta_c(x,y) = \frac{a_R\,(b_I(y) - f_I(x)) - a_I\,(b_R(y) - f_R(x))}{a_R^2 + a_I^2}. \tag{6.47}$$

These equations are well posed, and hence k, k_c are twice continuously differentiable functions of (x, y) (in particular, this fact ensures that the transformation (6.32), (6.33) is invertible). Equations (6.40)–(6.47) can be solved numerically using finite-dimensional discretization or with a symbolic recursive formula similar to (4.22)–(4.24).

The question remains of how to select f_R and f_I so that the target system (6.35)–(6.37) is exponentially stable. Let $c > 0$, and

$$\sup_{x\in[0,1]}\left(f_R(x) + \frac{1}{2}\left|f_I'(x)\right|\right) \le -\frac{c}{2}. \tag{6.48}$$

Consider the Lyapunov function

$$E(t) = \frac{1}{2}\int_0^1 \left(\tilde\rho(x,t)^2 + \tilde\imath(x,t)^2\right) dx. \tag{6.49}$$

Its time derivative along solutions of (6.35)–(6.37) is

$$\begin{aligned}
\dot E(t) &= \int_0^1 [\tilde\rho\,(a_R\tilde\rho_{xx} + f_R(x)\,\tilde\rho - a_I\tilde\imath_{xx} - f_I(x)\,\tilde\imath) \\
&\qquad + \tilde\imath\,(a_I\tilde\rho_{xx} + f_I(x)\,\tilde\rho + a_R\tilde\imath_{xx} + f_R(x)\,\tilde\imath)]\,dx \\
&= \int_0^1 (\tilde\rho\,(a_R\tilde\rho_{xx} + f_R(x)\,\tilde\rho - a_I\tilde\imath_{xx}) + \tilde\imath\,(a_I\tilde\rho_{xx} + a_R\tilde\imath_{xx} + f_R(x)\,\tilde\imath))\,dx \\
&= -\int_0^1 a_R\left(\tilde\rho_x^2 + \tilde\imath_x^2\right) dx + \int_0^1 f_R(x)\left(\tilde\rho^2 + \tilde\imath^2\right) dx + a_I\int_0^1 (\tilde\rho_x\tilde\imath_x - \tilde\imath_x\tilde\rho_x)\,dx \\
&\le \int_0^1 f_R(x)\left(\tilde\rho^2 + \tilde\imath^2\right) dx.
\end{aligned} \tag{6.50}$$

Hence, from (6.48) and the comparison principle, we have

$$E(t) \le E(0)\,e^{-ct} \qquad \text{for } t \ge 0. \tag{6.51}$$

To show that the state of the target system exponentially converges to zero for all $x \in [0, 1]$, we let

$$V(t) = \frac{1}{2}\int_0^1 \left(\tilde\rho_x^2(x,t) + \tilde\imath_x^2(x,t)\right) dx. \tag{6.52}$$

The time derivative of $V(t)$ along the solutions of (6.35)–(6.37) is

$$
\begin{aligned}
\dot V(t) &= \int_0^1 (\tilde\rho_x\tilde\rho_{xt} + \tilde\iota_x\tilde\iota_{xt})\,dx \\
&= -\int_0^1 (\tilde\rho_{xx}\tilde\rho_t + \tilde\iota_{xx}\tilde\iota_t)\,dx \\
&= -\int_0^1 [\tilde\rho_{xx}(a_R\tilde\rho_{xx} + f_R(x)\tilde\rho - a_I\tilde\iota_{xx} - f_I(x)\tilde\iota) \\
&\qquad + \tilde\iota_{xx}(a_I\tilde\rho_{xx} + f_I(x)\tilde\rho + a_R\tilde\iota_{xx} + f_R(x)\tilde\iota)]\,dx \\
&= -a_R\int_0^1 (\tilde\rho_{xx}^2 + \tilde\iota_{xx}^2)\,dx + \int_0^1 f_R(x)(\tilde\rho_x^2 + \tilde\iota_x^2)\,dx \\
&\quad + \int_0^1 f_I'(x)(\tilde\iota_x\tilde\rho - \tilde\rho_x\tilde\iota)\,dx - \frac{1}{2}\int_0^1 f_R''(x)(\tilde\rho^2 + \tilde\iota^2)\,dx \\
&\le \int_0^1 f_R(x)(\tilde\rho_x^2 + \tilde\iota_x^2)\,dx + \int_0^1 f_I'(x)(\tilde\iota_x\tilde\rho - \tilde\rho_x\tilde\iota)\,dx - \frac{1}{2}\int_0^1 f_R''(x)(\tilde\rho^2 + \tilde\iota^2)\,dx \\
&\le \int_0^1\left(f_R(x) + \frac{1}{2}\left|f_I'(x)\right|\right)(\tilde\rho_x^2 + \tilde\iota_x^2)\,dx + \frac{1}{2}\int_0^1 \left(\left|f_I'(x)\right| - f_R''(x)\right)(\tilde\rho^2 + \tilde\iota^2)\,dx \\
&\le \int_0^1\left(f_R(x) + \frac{1}{2}\left|f_I'(x)\right|\right)(\tilde\rho_x^2 + \tilde\iota_x^2)\,dx + \frac{1}{2}c_2\int_0^1 (\tilde\rho^2 + \tilde\iota^2)\,dx, \\
&\le -\frac{c}{2}V(t) + c_2E(0)\,e^{-ct},
\end{aligned} \tag{6.53}
$$

where we have used (6.48) and defined

$$
c_2 \triangleq \max\left\{\sup_{x\in[0,1]}\left(\left|f_I'(x)\right| - f_R''(x)\right), 0\right\}. \tag{6.54}
$$

From the comparison principle, we get

$$
V(t) \le \left(V(0) + 2\frac{c_2}{c}E(0)\right)e^{-\frac{c}{2}t} - 2\frac{c_2}{c}E(0)\,e^{-ct}, \tag{6.55}
$$

and hence we obtain

$$
V(t) \le \left(V(0) + \frac{2c_2}{c}E(0)\right)e^{-\frac{c}{2}t} \quad \text{for } t \ge 0. \tag{6.56}
$$

Since, by the Poincaré inequality,

$$
E(t) \le \frac{1}{2}V(t), \tag{6.57}
$$

we get

$$
V(t) \le c_3V(0)\,e^{-\frac{c}{2}t} \tag{6.58}
$$

with $c_3 = 1 + c_2/c$, which proves that the target system (6.35)–(6.37) is exponentially stable in H_1, provided that (6.48) is satisfied.

The particular choices of f_R and f_I that satisfy (6.48) have to be made carefully to avoid unnecessarily large control gains. This is achieved by avoiding the complete cancellation of the terms involving b_R and b_I in (6.23) by choosing f_R and f_I that cancel b_R and b_I only in the part of the domain $[x_s, 1]$ and that preserve the natural damping that exists in the plant downstream of x_s (in $[x_d, x_s]$). It ensures that only cancellation/domination of the source of instability is performed in the design, and the already stable part is kept unchanged. One possible good choice is

$$f_R(\breve{x}) = \begin{cases} -\frac{1}{2}c - \frac{1}{2}\left|b_I'(\breve{x})\right| \text{ for } x_s < x \le 1\,, \\ b_R(\breve{x}) \text{ for } \breve{x} \le x_s\,, \end{cases} \tag{6.59}$$

$$f_I(\breve{x}) = b_I(\breve{x}) \text{ for all } \breve{x}\,. \tag{6.60}$$

where x_s is chosen such that $f_R(x_s) + \frac{1}{2}\left|f_I'(x_s)\right| = -\frac{c}{2}$.

6.2.3 Observer Design

Since we consider the collocated actuator and sensor, and the control inputs are already chosen as $\rho(1,t), \iota(1,t)$, this leaves $y_R(t) = \rho_x(1,t)$ and $y_I(t) = \iota_x(1,t)$ for measurement.

The observer is designed along the lines of Chapter 5 and consists of a copy of the plant plus output injection terms both in the domain and in the boundary condition at $x = 1$:

$$\begin{aligned} \hat{\rho}_t &= a_R\hat{\rho}_{xx} + b_R(x)\hat{\rho} - a_I\hat{\iota}_{xx} - b_I(x)\hat{\iota} \\ &\quad + p_1(x)\left(y_R - \hat{y}_R\right) + p_{c,1}(x)\left(y_I - \hat{y}_I\right)\,, \end{aligned} \tag{6.61}$$

$$\begin{aligned} \hat{\iota}_t &= a_I\hat{\rho}_{xx} + b_I(x)\hat{\rho} + a_R\hat{\iota}_{xx} + b_R(x)\hat{\iota} \\ &\quad - p_{c,1}(x)\left(y_R - \hat{y}_R\right) + p_1(x)\left(y_I - \hat{y}_I\right) \end{aligned} \tag{6.62}$$

for $x \in (0,1)$, with boundary conditions $\hat{\rho}(0) = \hat{\iota}(0) = 0$ and

$$\hat{\rho}_x(1) = p_0(y_R - \hat{y}_R) + p_{c,0}(y_I - \hat{y}_I) + \int_0^1 \left[k(1,y)\hat{\rho}(y,t) + k_c(1,y)\hat{\iota}(y,t)\right]dy\,, \tag{6.63}$$

$$\hat{\iota}_x(1) = p_0(y_I - \hat{y}_I) - p_{c,0}(y_R - \hat{y}_R) + \int_0^1 \left[-k_c(1,y)\hat{\rho}(y,t) + k(1,y)\hat{\iota}(y,t)\right]dy\,. \tag{6.64}$$

In (6.61)–(6.64), $\hat{y}_R(t) = \hat{\rho}_x(1,t)$, $\hat{y}_I(t) = \hat{\iota}_x(1,t)$ and $p_1(x)$, $p_{c,1}(x)$, p_0, and $p_{c,0}$ are output injection gains to be designed. We omit the derivation of those gains and simply state the result:

$$p_1(x) = a_R k(1,x) + a_I k_c(1,x)\,, \tag{6.65}$$

$$p_{c,1}(x) = -a_I k(1,x) + a_R k_c(1,x)\,, \tag{6.66}$$

$$p_0 = 0\,, \tag{6.67}$$

$$p_{c,0} = 0\,. \tag{6.68}$$

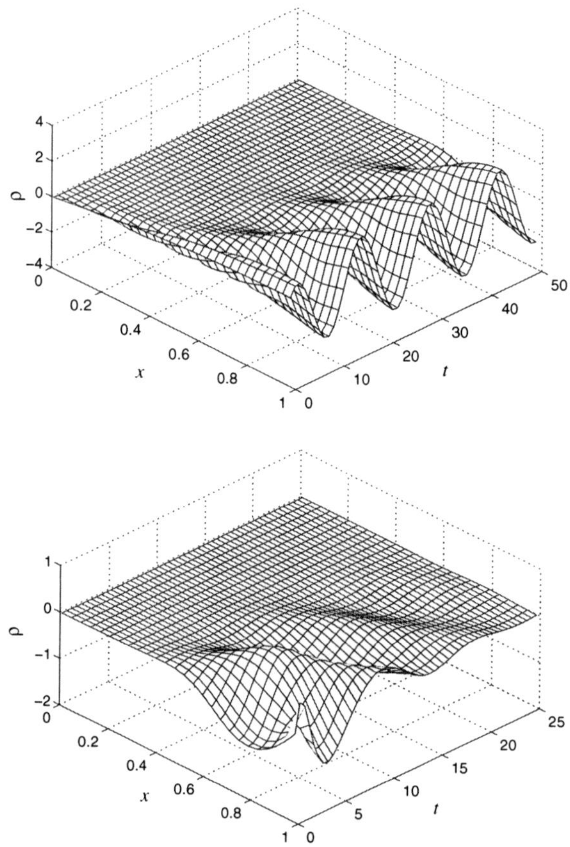

Figure 6.4. *Top: Open-loop simulation of the nonlinear plant. Bottom: Closed-loop response.*

The observer gains are expressed through control gains as a consequence of duality between the observer and control designs noted in Chapter 5.

6.2.4 Simulations

The controller and the observer can be combined into the output-feedback compensator. The proof of stability of the closed-loop system ("certainty equivalence") follows along the lines of Section 5.2.

The top graph in Figure 6.4 shows the open-loop (nonlinear) plant response for $x_d = -7$, at Reynolds number $Re = 60$.[9] (Only ρ is shown; ι looks qualitatively the same.) The system is linearly unstable and goes into a quasi-steady/limit-cycling motion reminiscent of vortex shedding. The feedback gains $k(1, x)$ and $k_c(1, x)$, as well as the observer gains $p_1(x)$ and $p_{c,1}(x)$, are shown in Figure 6.5. In Figure 6.6 the Bode plots of the compensator

[9]Defined as $Re = \rho U_\infty D/\mu$, where U_∞ is the free stream velocity, D is the cylinder diameter, and ρ and μ are density and viscosity of the fluid, respectively. Vortex shedding occurs when $Re > 47$.

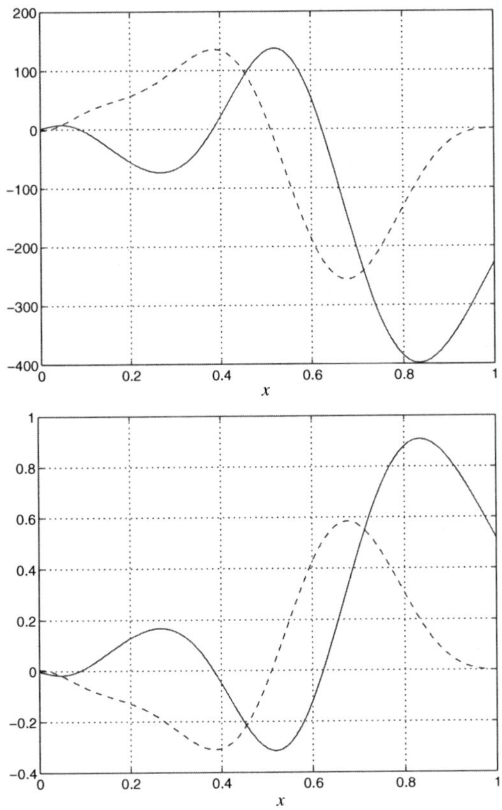

Figure 6.5. *Top: Gain kernels* $k(1,x)$ *(solid line) and* $k_c(1,x)$ *(dashed line). Bottom: Observer gains* $p_1(x)$ *(solid line) and* $p_{c,1}(x)$ *(dashed line).*

transfer functions are shown. The compensator is two-input–two-output; however, due to the symmetry in the plant, only two of the four transfer functions are different. The closed-loop response is shown in Figure 6.4 (top graph). We can see that vortex shedding is suppressed.

6.3 Notes and References

For more details on "passive damper" feedbacks for the Schrödinger equation, see [125, 67]. Exact controllability and observability results can be found in [108, 139, 124].

Modeling aspects of flows past bluff bodies are discussed in [110, 74, 148]. Numerical investigations based on the Navier–Stokes equations are numerous; see, for instance, [137, 65, 70]. Optimal controllers for the Ginzburg–Landau model have been designed for finite-dimensional approximations of equation (6.16) in [111] and [110]; the latter reference also provides an overview of previous work on stabilization of bluff body flows. In [2] globally stabilizing controllers were designed for a discretized nonlinear model.

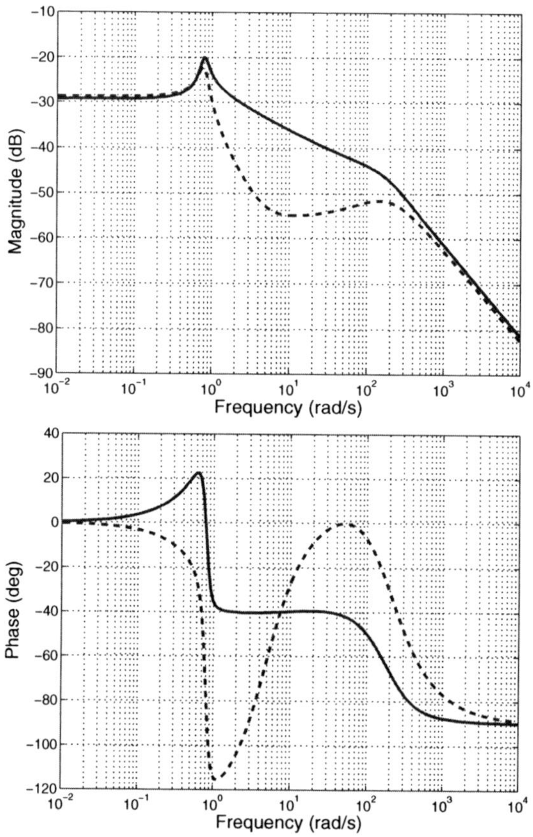

Figure 6.6. *Compensator transfer functions from $\rho_x(1)$ to $\rho(1)$ (solid lines) and $\iota_x(1)$ to $\rho(1)$ (dashed lines).*

Exercises

6.1. Prove the exponential stability of (6.5)–(6.7) with the help of the Lyapunov function

$$V = \frac{1}{2}\int_0^1 |w(y)|^2\, dy\,. \tag{6.69}$$

Note that $c > 0$.

6.2. Consider the Schrödinger equation

$$v_t = -j v_{xx}\,, \tag{6.70}$$
$$v_x(0) = -jqv(0) \tag{6.71}$$

with $q > 0$, which is unstable in the uncontrolled case $v(1) = 0$. Show that the change of variable

$$w(x) = v(x) + j(c_0 + q)\int_0^x e^{jq(x-y)} v(y)\,dy \tag{6.72}$$

and the boundary feedback

$$v(1) = -j(c_0 + q)\int_0^1 e^{jq(1-y)} v(y)\,dy \tag{6.73}$$

convert the closed-loop system into

$$w_t = -jw_{xx}\,, \tag{6.74}$$
$$w_x(0) = jc_0 w(0)\,, \tag{6.75}$$
$$w(1) = 0\,, \tag{6.76}$$

which happens to be exponentially stable for $c_0 > 0$. Note that this control law is designed to counter a source of instability at $x = 0$, using actuation at $x = 1$. (A similar design is considered in Exercise 7.3 for a wave/string equation.)

Chapter 7

Hyperbolic PDEs: Wave Equations

The material covered in the book thus far has dealt with parabolic PDEs—the heat equation, the reaction-advection-diffusion equations, and their more complicated varieties with $u(0,t)$ and $\int_0^x f(x,y)u(y,t)dy$ terms.

In this chapter we introduce the main ideas for backstepping control of hyperbolic PDEs, the most basic of which is the wave equation that models various oscillatory phenomena, i.e., string vibrations.

The main distinguishing feature of a wave equation is that it is second order in time. While the heat equation is

$$u_t = u_{xx}\,, \tag{7.1}$$

the wave equation is

$$u_{tt} = u_{xx}\,. \tag{7.2}$$

The difference is roughly analogous to that between the first-order ODE $\dot{z} + z = 0$ and a second-order ODE $\ddot{z} + z = 0$ and translates into a substantial difference in system behavior. While the solutions of the former decay monotonically to zero, the solutions of the latter oscillate. Of course, both heat equations and wave equations can be affected by additional phenomena which cause instability (instead of a monotonic decay or pure oscillation) but the key distinction remains: heat equation-type systems (with one derivative in time and two derivatives in space) have mostly real eigenvalues, whereas wave equation-type systems (with two derivatives in time and space) have mostly imaginary eigenvalues.

This heuristic discussion should help readers to understand the context in which control problems for the two classes of PDE systems have developed. While control systems for parabolic PDEs are mostly developed to accelerate their sluggish transient behavior, or to prevent exponential instability in the case of some chemically reacting or turbulent fluid flow systems, the control systems for hyperbolic PDEs are typically developed to damp out their oscillations (although situations also exist where control is also required to deal with exponential instabilities in wave equation systems; see Exercise 7.3).

The backstepping control ideas extend from parabolic to hyperbolic PDEs in a manner that is mathematically straightforward but conceptually rather tricky. Consider the elementary heat equation,

$$u_t = u_{xx}\,.$$

As we have seen in Chapter 4, a change of variable and boundary control can be designed to make the closed-loop system behave as

$$w_t = w_{xx} - cw\,,$$

with an arbitrary amount of in-domain damping $c > 0$. For the elementary wave equation,

$$u_{tt} = u_{xx}\,,$$

it would be highly desirable to design a variable change and boundary control to achieve closed-loop behavior governed by

$$w_{tt} = w_{xx} - cw_t\,,$$

where the terms $-cw_t$ represents the in-domain "viscous" damping. Unfortunately, this is not possible. As we shall see in this chapter, for wave equations it is necessary to pursue more subtle forms of adding damping, i.e., through special boundary conditions.

Many of the aspects of the analysis that we have encountered so far for parabolic PDEs become more delicate for hyperbolic PDEs. One aspect is the eigenvalue analysis—we devote some time in this chapter to the complex interplay between the boundary conditions and the resulting eigenvalues. More important, the Lyapunov analysis for wave equations is much more involved than it is for heat equations. The natural choice for a Lyapunov function is the total energy of the wave equation system, which consists of a sum of the potential and kinetic energy. This means that a Lyapunov function will involve the spatial L_2-norm both of the spatial derivative of the system state and of the time derivative of the system state—a much more complex situation than for parabolic PDEs, where the L_2-norm of the state itself (with no derivatives) was a good choice for a Lyapunov function.

The bulk of this chapter deals with the basic, undamped wave equation. However, we also discuss a more realistic form of the wave equation that includes a small amount of "Kelvin–Voigt damping." The Kelvin–Voigt damping models the internal material damping, which is present in any application, whether it be structural (strings, beams) or acoustic. Kelvin–Voigt damping prevents the system from having infinitely many eigenvalues on the imaginary axis which extend all the way to $\pm j\infty$. The presence of such eigenvalues in completely undamped wave equations is not only physically unrealistic but also a cause of certain artificial robustness considerations in the control design, as explained in Section 7.4. It is therefore very important to understand the control designs for wave equations with Kelvin–Voigt damping, although the study of the undamped wave equation is a key first step for understanding the peculiarities that arise in the control design and analysis for wave PDEs.

7.1 Classical Boundary Damping/Passive Absorber Control

The PDE that describes a vibrating string on a finite interval is

$$u_{tt} = u_{xx}\,, \tag{7.3}$$

$$u_x(0) = 0\,, \tag{7.4}$$

$$u(1) = 0\,. \tag{7.5}$$

The boundary conditions (7.4) and (7.5) correspond to the situation where the end of the string at $x = 1$ is "pinned" and the end $x = 0$ is "free." (The zero-slope boundary condition at $x = 0$ has the physical meaning of no force being applied at that end.)

To analyze stability properties of the solutions to this equation, consider the Lyapunov function

$$E = \frac{1}{2}\|u_x\|^2 + \frac{1}{2}\|u_t\|^2, \tag{7.6}$$

which represents the system energy at time t. The first term is the potential energy (u_x is shear) and the second term is kinetic energy (u_t is velocity). Taking the time derivative gives

$$\begin{aligned}
\dot{E} &= \int_0^1 u_x u_{xt} dx + \int_0^1 u_t u_{tt} dx \quad \text{(chain rule)} \\
&= \int_0^1 u_x u_{xt} dx + \int_0^1 u_t u_{xx} dx \\
&= \int_0^1 u_x u_{xt} dx + (u_t(x)u_x(x))|_0^1 - \int_0^1 u_{tx} u_x dx \quad \text{(integration by parts)} \\
&= (u_t(x)u_x(x))|_0^1 \\
&= 0\,.
\end{aligned}$$

Thus, $E(t) = E(0)$ and energy is conserved. In other words, the system is marginally stable. Of course, this is what is expected of an undamped string.

The classical method of stabilizing this system is to add a damping term to the boundary. Specifically, the boundary condition (7.4) is replaced by

$$u_x(0) = c_0 u_t(0). \tag{7.7}$$

The physical meaning of this boundary condition is that the force applied to the free end of the string is proportional to the velocity of the free end, which can be achieved using a passive damper/absorber.

Instead of using the Lyapunov/energy method to determine stability, we look at the eigenvalues of the system. First, the solution to (7.3) is sought in the form

$$u(x,t) = e^{\sigma t}\phi(x)\,.$$

Substituting this expression into (7.3) gives

$$\sigma^2 e^{\sigma t}\phi(x) = e^{\sigma t}\phi''(x)\,,$$

and using the boundary conditions, we get

$$e^{\sigma t}\phi(1) = 0\,,$$

$$e^{\sigma t}\phi'(0) = c_0\sigma e^{\sigma t}\phi(0)\,.$$

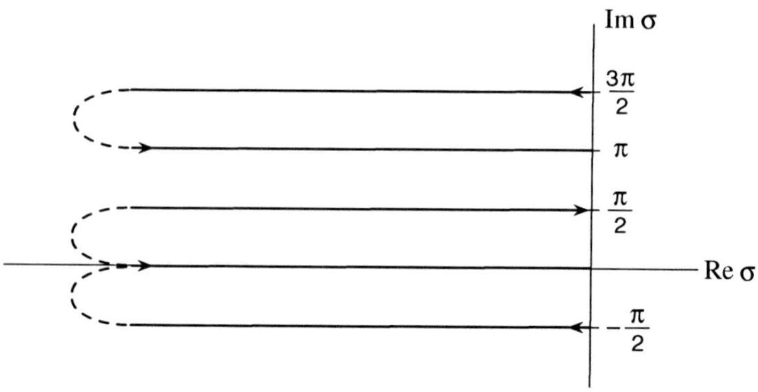

Figure 7.1. *The locus of eigenvalues of the system* (7.3), (7.7), (7.5) *when* c_0 *grows from* 0 *to* 1, *and then beyond* 1.

We have now arrived at the Sturm–Louiville problem

$$\phi'' - \sigma^2\phi = 0\,, \tag{7.8}$$
$$\phi'(0) = c_0\sigma\phi(0)\,, \tag{7.9}$$
$$\phi(1) = 0\,. \tag{7.10}$$

The solution of (7.8) is given by

$$\phi(x) = e^{\sigma x} + Be^{-\sigma x}\,. \tag{7.11}$$

From (7.10) we have $B = -e^{2\sigma}$. From (7.9),

$$\phi'(0) - c_0\sigma\phi(0) = 0\,, \tag{7.12}$$
$$\sigma(1 + e^{2\sigma}) - c_0\sigma(1 - e^{2\sigma}) = 0\,, \tag{7.13}$$
$$e^{2\sigma} = -\frac{1 - c_0}{1 + c_0}\,. \tag{7.14}$$

Solving for σ gives

$$\sigma = -\frac{1}{2}\ln\left|\frac{1 + c_0}{1 - c_0}\right| + j\pi\begin{cases} n + \frac{1}{2} & 0 \le c_0 < 1\,, \\ n & c_0 > 1\,, \end{cases} \tag{7.15}$$

and the solution of the system converges to zero in finite time for $c_0 = 1$. In Figure 7.1 we can see the locus of eigenvalues (7.15) in the complex plane as the value of c_0 changes.

The boundary control provided by (7.7) has the capacity to add considerable damping to the system; however, it requires actuation on the free end $x = 0$, which is not always feasible.

7.2 Backstepping Design: A String with One Free End and Actuation on the Other End

Let us now consider the wave equation with damping boundary control at the constrained end:

$$u_{tt} = u_{xx}\,, \tag{7.16}$$
$$u_x(0) = 0\,, \tag{7.17}$$
$$u_x(1) = -c_1 u_t(1)\,, \tag{7.18}$$

where $c_1 > 0$. This system has an arbitrary constant as an equilibrium profile. To deal with this multitude of equilibria, we need a more sophisticated controller at $x = 1$ (backstepping) if the boundary condition at $x = 0$ is to remain free. We will now design such a controller using the backstepping approach.

We propose the transformation

$$w(x) = u(x) + c_0 \int_0^x u(y)\,dy\,, \tag{7.19}$$

which maps the plant into the target system

$$w_{tt} = w_{xx}\,, \tag{7.20}$$
$$w_x(0) = c_0 w(0)\,, \tag{7.21}$$
$$w_x(1) = -c_1 w_t(1)\,. \tag{7.22}$$

The idea is that a large c_0 in the boundary condition at $x = 0$ can make $w_x(0) = c_0 w(0)$ behave as $w(0) = 0$. (Large c_0 is not necessary; it is a design option, and hence one should not view this as employing high gain feedback.)

First, we need to analyze the stability of the target system. Consider the Lyapunov function

$$V = \frac{1}{2}\left(\|w_x\|^2 + \|w_t\|^2 + c_0 w^2(0)\right) + \delta \int_0^1 (1+x) w_x(x) w_t(x) dx\,. \tag{7.23}$$

The crucial novelty relative to the previous Lyapunov functions used in this book is the introduction of an "indefinite" spatially weighted cross-term between w_x and w_t. Using the Cauchy–Schwarz and Young's inequalities, one can show that for sufficiently small δ there exist $m_1,\ m_2 > 0$ such that

$$m_1 U \leqslant V \leqslant m_2 U, \quad U = \|w_x\|^2 + \|w_t\|^2 + w^2(0)\,. \tag{7.24}$$

Therefore V is positive definite.

The derivative of V along the solution of the target system is

$$\begin{aligned}
\dot V &= \int_0^1 w_x w_{tx} dx + \int_0^1 w_t w_{tt} dx + c_0 w(0) w_t(0) \\
&\quad + \delta \int_0^1 (1+x)(w_{xt} w_t + w_x w_{tt}) dx \\
&= \int_0^1 w_x w_{tx} dx \\
&\quad + \int_0^1 w_t w_{xx} dx + w_x(0) w_t(0) + \delta \int_0^1 (1+x)(w_{xt} w_t + w_x w_{xx}) dx \\
&= \int_0^1 w_x w_{tx} dx + w_t w_x |_0^1 - \int_0^1 w_t w_{xt} dx + w_x(0) w_t(0) \\
&\quad + \delta \int_0^1 (1+x)(w_{xt} w_t + w_x w_{xx}) dx \\
&= \delta \left(\int_0^1 w_{xt} w_t dx + \int_0^1 w_x w_{xx} dx + \int_0^1 x w_{xt} w_t dx + \int_0^1 x w_x w_{xx} dx \right) \\
&\quad + w_t(1) w_x(1) \,.
\end{aligned}$$

In the last two integrals we notice that

$$w_{xt} w_t dx = \frac{d}{dx}\frac{w_t^2}{2}, \qquad w_x w_{xx} dx = \frac{d}{dx}\frac{w_x^2}{2} \tag{7.25}$$

and use integration by parts:

$$\begin{aligned}
\dot V &= w_t(1) w_x(1) + \frac{\delta}{2}\left[(1+x)(w_x^2 + w_t^2)\right]|_0^1 - \frac{\delta}{2}\left[\|w_x\|^2 + \|w_t\|^2\right] \\
&= -c_1 w_t^2 + \delta(w_t^2(1) + w_x^2(1)) - \frac{\delta}{2}\left[w_x^2(0) + w_t^2(0)\right] - \frac{\delta}{2}\left[\|w_x\|^2 + \|w_t\|^2\right] \\
&= -\left(c_1 - \delta(1+c_1^2)\right) w_t^2(1) - \frac{\delta}{2}\left(w_t^2(0) + c_0^2 w^2(0)\right) - \frac{\delta}{2}\left[\|w_x\|^2 + \|w_t\|^2\right],
\end{aligned}$$

which is negative definite for $\delta < \frac{c_1}{1+c_1^2}$. Since δ is just an analysis parameter, we can choose it to be arbitrarily small. It now follows from (7.23) and (7.24) that

$$U(t) \le M e^{-t/M} U(0)$$

for some possibly large M, which proves the exponential stability of the target system.

The resulting Neumann backstepping controller is obtained by differentiating the transformation (7.19) and setting $x = 1$:

$$u_x(1) = -c_1 u_t(1) - c_0 u(1) - c_1 c_0 \int_0^1 u_t(y) dy \,. \tag{7.26}$$

For the best performance (with respect to the case with a "pinned" uncontrolled boundary condition in Section 7.1), one should choose c_0 to be large and c_1 approximately 1. Some insight into the properties of the backstepping controller may be obtained by examining the individual terms in (7.26). The term $-c_1 u_t(1)$ provides boundary damping so that the action of $-c_1 u_t(1) - c_0 u(1)$ is similar to PD control. The last term is a spatially averaged velocity and is the backstepping term that allows actuation at the constrained end $u_x(1)$.

The corresponding output-feedback controller with only boundary sensing is

$$u_x(1) = -c_0 \hat{u}(1) - c_1 \hat{u}_t(1) + c_0 c_1 \int_0^1 \hat{u}_t(y) dy \,, \tag{7.27}$$

where the observer state is governed by

$$\hat{u}_{tt} = \hat{u}_{xx} \,, \tag{7.28}$$

$$\hat{u}_x(0) = \tilde{c}_0(\hat{u}_t(0) - u_t(0)) \,, \tag{7.29}$$

$$\hat{u}(1) = u(1) \tag{7.30}$$

with $\tilde{c}_0 > 0$.

7.3 Wave Equation with Kelvin–Voigt Damping

An entirely different backstepping design is possible when the wave equation has a small amount of Kelvin–Voigt damping (internal material damping, present in all realistic materials):

$$u_{tt} = u_{xx} + d u_{xxt} \,, \tag{7.31}$$

$$u_x(0) = 0 \,, \tag{7.32}$$

$$u(1) = \text{control}, \tag{7.33}$$

where d is a small positive constant.

We use the transformation

$$w(x) = u(x) - \int_0^x k(x, y) u(y) \, dy$$

to transform the original system into the target system

$$w_{tt} = (1 + d\partial_t)(w_{xx} - cw) \,, \tag{7.34}$$

$$w_x(0) = 0 \,, \tag{7.35}$$

$$w(1) = 0 \,. \tag{7.36}$$

The nth pair of eigenvalues σ_n of this system satisfies the quadratic equation

$$\sigma_n^2 + d\left[c + \left(\frac{\pi}{2} + \pi n\right)^2\right]\sigma_n + \left[c + \left(\frac{\pi}{2} + \pi n\right)^2\right] = 0 \,, \tag{7.37}$$

where $n = 0, 1, 2, \ldots$. There are two sets of eigenvalues: For lower n the eigenvalues reside on the circle

$$\left(\mathrm{Re}(\sigma_n) + \frac{1}{d}\right)^2 + (\mathrm{Im}(\sigma_n))^2 = \frac{1}{d^2}, \tag{7.38}$$

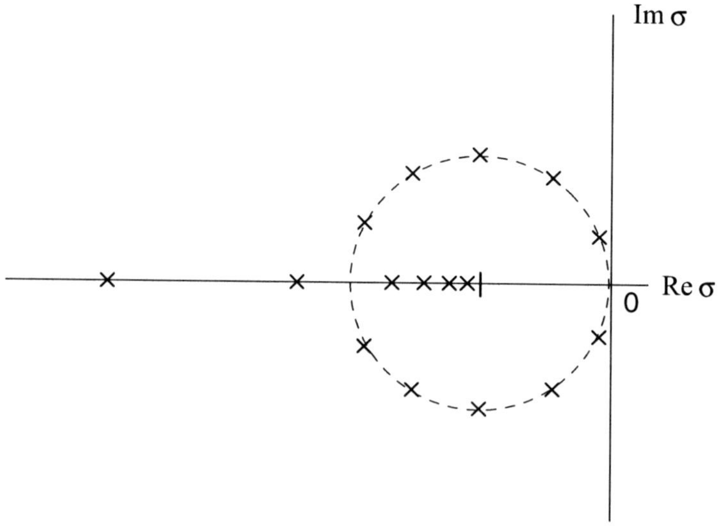

Figure 7.2. *Open-loop eigenvalues for the wave equation with Kelvin–Voigt damping.*

and for higher n the eigenvalues are real, with one branch accumulating towards $-1/d$ as $n \to \infty$ and the other branch converging to $-\infty$. The open-loop eigenvalues ($c = 0$) are shown in Figure 7.2. Increasing c moves the eigenvalues along the circle in the negative real direction and decreases the number of them on the circle (ultimately they become real). With a very high value of c all of the eigenvalues can be made real. While possible, this would not necessarily be a good idea, not for transient response or for disturbance attenuation, and certainly not from the point of view of control effort. Thus, the flexibility of improving the damping using the backstepping transformation and controller should be used judiciously, with lower values of c if d is already relatively high.

The kernel PDE can be shown as

$$k_{xx} = k_{yy} + ck\,, \tag{7.39}$$

$$k_y(x,0) = 0\,, \tag{7.40}$$

$$k(x,x) = \frac{c}{2}x\,. \tag{7.41}$$

Note that this is the same PDE as in Example 4.1 from Chapter 4. Its solution is

$$k(x,y) = -cx\frac{I_1\left(\sqrt{c(x^2-y^2)}\right)}{\sqrt{c(x^2-y^2)}}\,. \tag{7.42}$$

The controller is given by

$$u(1) = -\int_0^1 c\frac{I_1\left(\sqrt{c(1-y^2)}\right)}{\sqrt{c(1-y^2)}}u(y)\,dy\,. \tag{7.43}$$

Note that the controller does not depend on d.

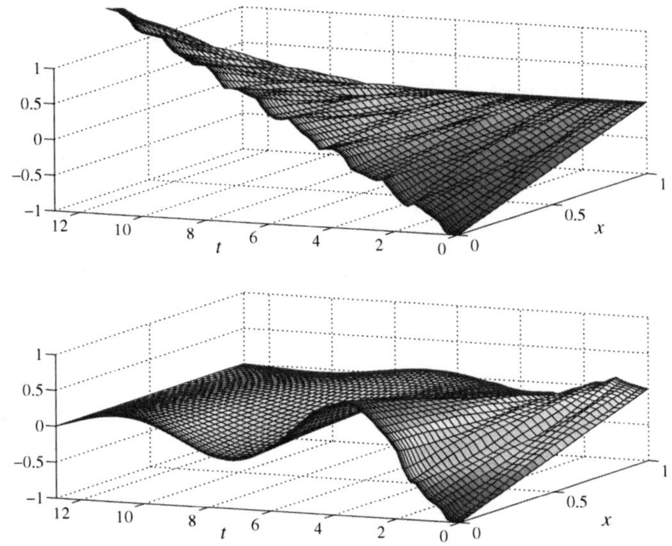

Figure 7.3. *Open-loop (top) and closed-loop (bottom) responses of the unstable wave equation from Exercise* 7.3.

7.4 Notes and References

The literature that covers control of wave equations—hundreds of papers and several books on controllability, stabilization, optimal control, and state and parameter estimation—is too vast to review here. We have introduced the ideas for backstepping control of wave equations in [96, 99].

It is important to comment on the robustness of boundary damper-based control laws (which includes the backstepping design). The question of robustness of such feedbacks to a time delay (due to measurement, computation, and actuation) received considerable attention in the 1980s and 1990s. It has been shown that such feedbacks have a zero robustness margin to the presence of a delay; namely, they are destabilized by any, infinitesimal amount of delay. This is in fact true of other more complicated feedbacks, though it is more difficult to establish analytically due to increased complexity. The reason for the lack of robustness, however, lies not in the control design but in the model itself. The undamped wave equation, with its infinitely many eigenvalues on the $j\omega$ axis, is simply not a physically realistic model. As soon as the undamped wave equation is augmented by a small amount of Kelvin–Voigt damping, which is present in any structure and evident in any experiment through a finite (rather than infinite) number of resonant frequencies in the frequency response or in the power spectral density, a robustness property of any experimentally stabilizing design to the presence of a *small* delay is restored.

It is also important to caution the reader that a wave equation with Kelvin–Voigt damping cannot be categorized as a *hyperbolic* PDE. With at most a finite number of conjugate-complex eigenvalues in its spectrum, such a PDE is a parabolic/hyperbolic hybrid.

Exercises

7.1. Show that the transformation

$$w(x) = u(x) + c_0 \int_0^x u(y)\,dy$$

and the boundary control

$$u_x(1) = -c_0 u(1) - c_1 \left(u_t(1) + c_0 \int_0^1 u_t(y)\,dy \right)$$

convert the plant

$$u_{tt} = u_{xx}\,,$$
$$u_x(0) = 0$$

into the asymptotically stable system

$$w_{tt} = w_{xx}\,,$$
$$w_x(0) = c_0 w(0)\,,$$
$$w_x(1) = -c_1 w_t(1)\,.$$

7.2. In Exercise 7.1 determine c_0 and c_1 such that the first pair of poles is approximately $-1.15 \pm j1.5$. In order to do this, use the program downloadable from `http://flyingv.ucsd.edu/krstic/teaching/287/gui_wave2.zip`.

7.3. Consider the wave equation

$$u_{tt} = u_{xx}\,,$$
$$u_x(0) = -qu(0)\,,$$

which is unstable with $u(1) = 0$ when $q \geq 1$. Show that the change of variable

$$w(x) = u(x) + (c_0 + q) \int_0^x e^{q(x-y)} u(y)\,dy$$

and the boundary feedback

$$u_x(1) = -c_1 u_t(1) - (c_0 + q)u(1) - (c_0 + q) \int_0^1 e^{q(1-y)} [c_1 u_t(y) + qu(y)] dy$$

convert the closed-loop system into

$$w_{tt} = w_{xx}\,,$$
$$w_x(0) = c_0 w(0)\,,$$
$$w_x(1) = -c_1 w_t(1)\,.$$

Note: In Figure 7.3 it is illustrated that the open-loop plant is unstable and that the feedback controller designed in this exercise successfully stabilizes the plant.

Chapter 8
Beam Equations

While the wave equation is the most appropriate "point of entry" into the realm of hyperbolic PDEs, beam equations are considered a physically relevant benchmark for control of hyperbolic PDEs and structural systems in general.

We start by discussing the main differences between wave equation (string) models,

$$u_{tt} - u_{xx} = 0\,, \tag{8.1}$$

$$u_x(0) = 0 \quad \text{(free end)}\,, \tag{8.2}$$

$$u(1) = 0 \quad \text{(pinned end)}\,, \tag{8.3}$$

and beam models. The simplest beam model is the Euler–Bernoulli model

$$u_{tt} + u_{xxxx} = 0\,, \tag{8.4}$$

$$u_{xx}(0) = u_{xxx}(0) = 0 \;\; \text{(free end condition)}\,, \tag{8.5}$$

$$u(0) = u_x(0) = 0 \quad \text{(clamped end condition)}. \tag{8.6}$$

The obvious difference between the PDEs (8.1) and (8.4) is in the number of spatial derivatives—the wave equation is second order in x, whereas the Euler–Bernoulli beam model is fourth order in x. One consequence of this difference is that a wave equation requires one boundary condition per end point (see (8.2) or (8.3)), whereas the Euler–Bernoulli beam model requires two boundary conditions per end point; see (8.5) or (8.6). A more important difference is in the eigenvalues. Both the beam and the string models have all of their eigenvalues on the imaginary axis. However, while the string eigenvalues are equidistant (growing linearly in n), the beam eigenvalues get further and further apart as they go up the $j\omega$ axis (they grow quadratically in n). This difference in the eigenvalue pattern is a consequence of the difference in the number of derivatives in x.

A reader might ask how these differences translate into control. Is it obvious that a beam is more difficult to control than a string? The answer is not clear and is not necessarily "yes." While the presence of higher derivatives clearly generates some additional issues to deal with in the control design, the wave equation has its own peculiarities that one should not underestimate. For example, controllability results for beams are valid on arbitrary short time intervals, whereas for strings such results hold only over time intervals that are lower bounded in proportion to the "wave propagation speed" of the string (which physically corresponds to "string tension"). Also, it is almost intuitively evident that keeping a string from vibrating may not be easier than keeping a beam from vibrating.

Beam modeling is not an elementary topic, even for slender beams undergoing only small displacements (in which case a linear single-PDE 1D model can be arrived at). The catalog of linear slender-beam models consists of the following four models:

(1) Euler–Bernoulli model,

(2) Rayleigh model,

(3) shear beam model, and

(4) Timoshenko model.

The Euler–Bernoulli model is the simplest and includes neither the effect of shear deformations nor rotary inertia, whereas the Timoshenko model includes both effects and is the most general and most complex model. (All of the models include the effects of lateral displacement and bending moment, with the former contributing the kinetic energy and the latter the strain/potential energy.)

The Rayleigh and shear beam models are mathematically identical although they are physically different (the parameters that appear in the two models are different). Both models include derivative terms not only of the forms u_{tt} and u_{xxxx} but also of the form u_{xxtt}.

The traditional way to stabilize the Euler–Bernoulli beam is similar to the passive damper design for the wave equation: Use a damping boundary feedback at the tip of the beam

$$u_{xx}(0) = c_0 u_{xt}(0), \quad c_0 > 0. \tag{8.7}$$

This design, while damping the higher modes, is not capable of adding a lot of damping to the first few modes. We present a design based on backstepping implemented by actuating at the base of the beam, which has the ability to add damping to all the modes and to even stabilize the beam in the case when it is open-loop unstable due to destabilizing forces acting at the tip. This design is presented in Section 8.2.

Before dealing with the Euler–Bernoulli beam model, we present a backstepping design for the seemingly more complex shear beam model. Ironically, the design for the shear beam is more accessible than that for the simpler, Euler–Bernoulli beam. The reason for this is that the additional term in the shear beam model, accounting for the finiteness of the shear modulus, has a particular "regularizing" effect on the system. In more precise terms, the shear beam model is essentially a partial integrodifferential equation (PIDE) of order two in x, whereas the Euler–Bernoulli beam is fundamentally of order four in x. This results in a form of methodological "discontinuity," where, even though in the "infinite-shear-modulus" limit the shear beam model becomes the Euler–Bernoulli model, this does not occur with the designs. The design for the shear beam is defined only when the shear modulus is finite (the control gains go to infinity as the shear modulus goes to infinity). The design for the Euler–Bernoulli model is fundamentally different from the design for the shear beam.

8.1 Shear Beam

8.1.1 Shear Beam Model

We consider the shear beam model, usually written as

$$u_{tt} - \varepsilon u_{xxtt} + u_{xxxx} = 0, \tag{8.8}$$

where ε is a small constant inversely proportional to the shear modulus (and unrelated to damping). We can rewrite this model in the form of the wave equation coupled with a second-order ODE in x as follows:

$$\varepsilon u_{tt} = u_{xx} - \alpha_x \tag{8.9}$$

$$0 = \varepsilon \alpha_{xx} - \alpha + u_x \,, \tag{8.10}$$

where α is the deflection angle due to the bending of the beam. The free-end boundary condition is given by

$$u_x(0) = \alpha(0) \,, \tag{8.11}$$

$$\alpha_x(0) = 0. \tag{8.12}$$

One can verify that this model is equivalent to the model (8.8) by following these steps:

(a) $(8.9)_x + (8.10) = (\star)$.
(b) $(\star)_x = (\star\star)$.
(c) $(\star\star) - \frac{1}{\varepsilon}(8.9) = (8.8)$.

8.1.2 Control Design

The first step of our design is to solve the ODE (8.10) for α:

$$\alpha(x) = \cosh(bx)\alpha(0) - b\int_0^x \sinh(b(x-y))u_y(y)dy \,, \tag{8.13}$$

where $b = 1/\sqrt{\varepsilon}$. This solution is easily obtained via Laplace transform in the spatial variable x.

The constant $\alpha(0)$ in (8.13) can be expressed in terms of $\alpha(1)$ in the following way:

$$\begin{aligned}\alpha(0) &= \frac{1}{\cosh(b)}\left[\alpha(1) + b\int_0^1 \sinh(b(1-y))u_y(y)dy\right] \\ &= \frac{1}{\cosh(b)}\left[\alpha(1) + b\sinh(b(1-y))u(y)|_0^1 + b^2\int_0^1 \cosh(b(1-y))u(y)dy\right] \\ &= \frac{1}{\cosh(b)}\left[\alpha(1) - b\sinh(b)u(0) + b^2\int_0^1 \cosh(b(1-y))u(y)dy\right]. \end{aligned} \tag{8.14}$$

The integral term on the right-hand side of this equality is not spatially causal, due to the fourth derivative in the original shear beam model (8.8). To put the system into a strict-feedback form, we eliminate this integral by choosing

$$\alpha(1) = b\sinh(b)u(0) - b^2\int_0^1 \cosh(b(1-y))u(y)dy \tag{8.15}$$

so that $\alpha(0) = 0$. Then we have

$$\alpha(x) = b\sinh(bx)u(0) - b^2 \int_0^x \cosh(b(x-y))u(y)dy\,. \tag{8.16}$$

Differentiating $\alpha(x)$ with respect to x and substituting the result into the wave equation (8.9), we get the system in the form ready for the control design:

$$\varepsilon u_{tt} = u_{xx} + b^2 u - b^2\cosh(bx)u(0) + b^3 \int_0^x \sinh(b(x-y))u(y)dy\,. \tag{8.17}$$

Following our procedure, we use the transformation

$$w(x) = u(x) - \int_0^x k(x,y)u(y)\,dy \tag{8.18}$$

to map the system (8.17) into the following exponentially stable target system:

$$\varepsilon w_{tt} = w_{xx}\,, \tag{8.19}$$

$$w_x(0) = c_0 w(0)\,, \tag{8.20}$$

$$w_x(1) = -c_1 w_t(1)\,, \tag{8.21}$$

where c_0 and c_1 are design parameters. It may be surprising that we use the same target system here as in the control design for the wave equation in the previous chapter. We discuss this choice in more detail in Section 8.3.

Substituting the transformation (8.18) into the target system, one can derive the following PDE for the kernel $k(x, y)$:

$$k_{xx} = k_{yy} + b^2 k - b^3\sinh(b(x-y)) + b^3 \int_y^x k(x,\xi)\sinh(b(\xi - y))d\xi\,, \tag{8.22}$$

$$k(x,x) = -\frac{b^2}{2}x - c_0\,, \tag{8.23}$$

$$k_y(x,0) = b^2 \int_0^x k(x,y)\cosh(by)dy - b^2\cosh(bx)\,. \tag{8.24}$$

This PDE has to be solved numerically.

The second boundary controller (the first one is given by (8.15)) is obtained by differentiating (8.18) with respect to x and setting $x = 1$:

$$u_x(1) = k(1,1)u(1) + \int_0^1 k_x(1,y)u(y)dy - c_1 u_t(1) + c_1 \int_0^1 k(1,y)u_t(y)dy\,. \tag{8.25}$$

In a similar way, one can design the observer to avoid needing of position and velocity measurements along the beam.

8.1.3 On the Choice of the Wave Equation as the Target System

As we indicated after (8.21), our choice of a wave equation (8.19) as a target system for shear beam stabilization may seem counterintuitive.

A reader may have concerns on two levels. First, mathematically, there seems to be a loss of two derivatives in x, from four in the original plant and to two in the target system. This is only a matter of appearance. Instead of the model (8.8) we use the representation (8.9)–(8.12), which can be written as a PIDE with two derivatives in x,

$$\begin{aligned}\varepsilon u_{tt} &= u_{xx} + b^2 u - b^2 \cosh(bx) u(0) + b^3 \int_0^x \sinh(b(x-y)) u(y)\, dy \\ &\quad - \frac{b \sinh(bx)}{\cosh(b)} \left[\alpha(1) - b \sinh(b) u(0) + b^2 \int_0^1 \cosh(b(1-y)) u(y)\, dy \right], \qquad (8.26)\\ u_x(0) &= \frac{1}{\cosh(b)} \left[\alpha(1) - b \sinh(b) u(0) + b^2 \int_0^1 \cosh(b(1-y)) u(y)\, dy \right]. \qquad (8.27)\end{aligned}$$

This is an explicit way of writing

$$\varepsilon u_{tt} = \left(\partial_{xx} - \frac{1}{\varepsilon} \right)^{-1} u_{xxxx} = \left(1 - \frac{1}{\varepsilon} \partial_{xx}^{-1} \right)^{-1} u_{xx}, \qquad (8.28)$$

which follows (8.8) and clearly displays a predominantly second-order character in x. When viewed in this way, namely, as a complicated wave equation in the form of (8.26) with integral terms and $u(0)$ terms, then the transformation into the "clean" wave equation format (8.19), with the variable change (8.18), (8.22)–(8.24) and the boundary controls (8.15), (8.25), is a perfectly natural choice. A reader's second possible concern about the transformation of a beam into the wave equation (8.19) may be based on purely physical grounds, i.e., that we may be forcing a rather rigid object (a beam) to act dynamically as a very flexible object (a string). This concern is also just a matter of appearance. By observing in (8.19) that ε is small, we realize that we are transforming our shear beam into a highly tensioned string, which is a reasonable thing to do.

8.1.4 Simulation Results

In Figures 8.1 and 8.2 we present the simulation results for the beam with the designed controller. From Figure 8.1 it is clear that our design for shear beam is very effective and physically reasonable, despite being based on the wave equation as a target system. The closed-loop damping performance is excellent, whereas the control effort, marked by the motion of the "dot" on the right side of the domain (in the set of plots on the right side of the figure) is very low. It is clear that the controller achieves excellent damping performance by using good "timing" of the control inputs rather than by relying on a large magnitude of the control inputs.

8.1.5 Extension to an Unstable Beam

The results in Figure 8.1 are actually for an unstable beam, which is being destabilized by a "repulsive" force at the free end, modeled by a boundary condition $u_x(0) = \alpha(0) - qu(0)$, $q > 0$, instead of by (8.11). This is similar to the unstable wave equation example from Exercise 7.3. The control applied is of the same form as (8.25), but the gain kernel is slightly different, with an extra term $-q$ in (8.23) and an extra term $-qk(x, 0)$ in (8.24).

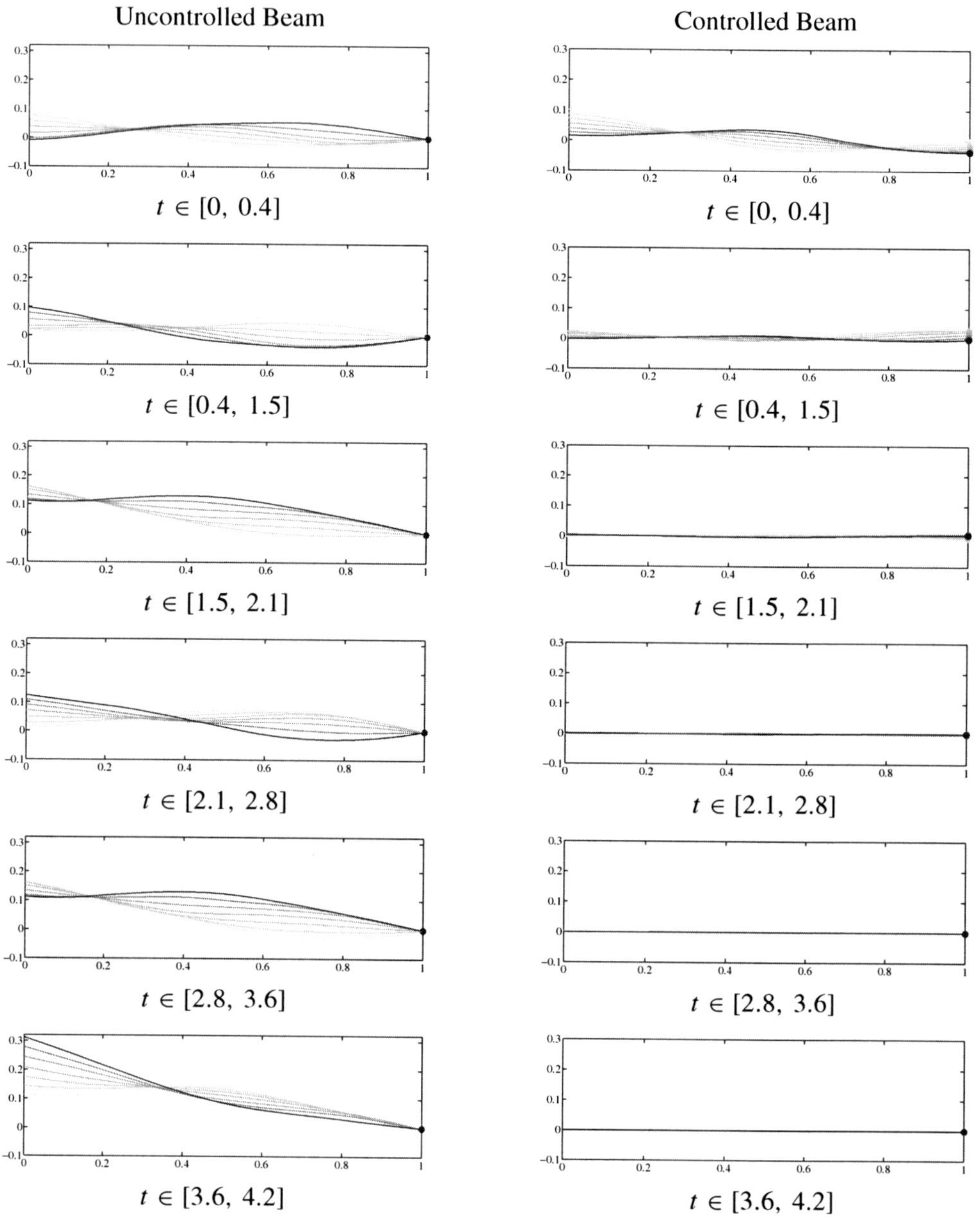

Figure 8.1. *Snapshots of the shear beam movements with increasing shading denoting increasing time in the sequences.*

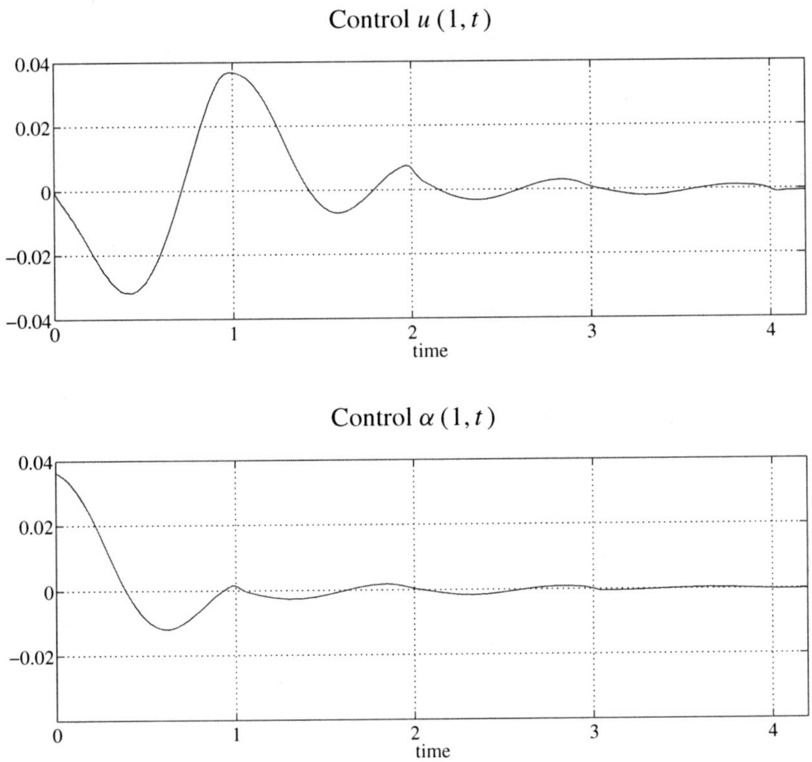

Figure 8.2. *The controls for the shear beam.*

8.2 Euler–Bernoulli Beam

Consider the Euler–Bernoulli beam model

$$u_{tt} + u_{xxxx} = 0 \tag{8.29}$$

with boundary conditions

$$u_x(0) = u_{xxx}(0) = 0\,, \tag{8.30}$$

which correspond to a "sliding end." We assume the following boundary conditions at $x = 1$:

$$u(1) = U_1, \quad u_{xx}(1) = U_2\,, \tag{8.31}$$

where U_1 and U_2 are control inputs. The open-loop case $U_1 = U_2 \equiv 0$ corresponds to a "pinned end."

To start the control design for (8.29)–(8.31), we introduce a new complex variable

$$v = u_t - ju_{xx}\,. \tag{8.32}$$

The direct substitution shows that v defined in this way satisfies the Schrödinger equation (6.1), (6.2). We can therefore apply the design (6.3)–(6.11) to the Euler–Bernoulli beam.

8.2.1 Gain Kernels

The gain kernel of the transformation (6.3) satisfies the PDE (6.8)–(6.10). Let

$$k(x,y) = r(x,y) + js(x,y)\,; \tag{8.33}$$

then we get two coupled PDEs for the gains $r(x, y)$ and $s(x, y)$:

$$r_{xx}(x,y) = r_{yy}(x,y) - cs(x,y)\,, \tag{8.34}$$
$$r_y(x,0) = 0\,, \tag{8.35}$$
$$r(x,x) = 0 \tag{8.36}$$

and

$$s_{xx}(x,y) = s_{yy}(x,y) + cr(x,y)\,, \tag{8.37}$$
$$s_y(x,0) = 0\,, \tag{8.38}$$
$$s(x,x) = -\frac{c}{2}x\,. \tag{8.39}$$

These PDEs can be decoupled into two fourth-order PDEs:

$$r_{xxxx}(x,y) - 2r_{xxyy}(x,y) + r_{yyyy}(x,y) = -c^2 r(x,y)\,, \tag{8.40}$$
$$r_y(x,0) = 0\,, \tag{8.41}$$
$$r_{yyy}(x,0) = 0\,, \tag{8.42}$$
$$r(x,x) = 0\,, \tag{8.43}$$
$$r_{xx}(x,x) - r_{yy}(x,x) = \frac{c^2}{2}x \tag{8.44}$$

and

$$s_{xxxx}(x,y) - 2s_{xxyy}(x,y) + s_{yyyy}(x,y) = c^2 s\,, \tag{8.45}$$
$$s_y(x,0) = 0\,, \tag{8.46}$$
$$s_{yyy}(x,0) = 0\,, \tag{8.47}$$
$$s(x,x) = -\frac{c}{2}x\,, \tag{8.48}$$
$$s_{xx}(x,x) - s_{yy}(x,x) = 0\,. \tag{8.49}$$

The solutions to these PDEs are obtained from the solution (6.11):

$$r(x,y) = x\sqrt{\frac{c}{2\left(x^2-y^2\right)}}\left[-\mathrm{ber}_1\left(\sqrt{c\left(x^2-y^2\right)}\right) - \mathrm{bei}_1\left(\sqrt{c\left(x^2-y^2\right)}\right)\right], \tag{8.50}$$

$$s(x,y) = x\sqrt{\frac{c}{2\left(x^2-y^2\right)}}\left[\mathrm{ber}_1\left(\sqrt{c\left(x^2-y^2\right)}\right) - \mathrm{bei}_1\left(\sqrt{c\left(x^2-y^2\right)}\right)\right]. \tag{8.51}$$

8.2.2 Target System

Let us define

$$\alpha(x) = \int_x^1 \int_0^y \operatorname{Im}\{w(\xi)\}\, d\xi dy\,, \tag{8.52}$$

where w is the state of the target system (6.5)–(6.7) for the Schrödinger equation. Then it is easy to verify that α satisfies the following fourth-order PDE:

$$\alpha_{tt} + 2c\alpha_t + c^2\alpha + \alpha_{xxxx} = 0 \tag{8.53}$$

with boundary conditions

$$\alpha_x(0) = 0\,, \tag{8.54}$$

$$\alpha_{xxx}(0) = 0\,, \tag{8.55}$$

$$\alpha_{xx}(1) = 0\,, \tag{8.56}$$

$$\alpha(1) = 0\,. \tag{8.57}$$

That this target system is exponentially stable is easily seen from the definition (8.52) and the fact that (6.5)–(6.7) is exponentially stable. With a straightforward computation we obtain the eigenvalues of (8.53)–(8.57):

$$\sigma_n = -c \pm j\frac{\pi^2}{4}(2n+1)^2 \quad \text{for } n = 0, 1, 2, \ldots. \tag{8.58}$$

8.2.3 Transformation and Control Laws

From (8.52) and (6.5)–(6.7), it follows that the state w is expressed through α in the following way:

$$w = \alpha_t + c\alpha - j\alpha_{xx}\,. \tag{8.59}$$

The transformation (6.3) becomes

$$\alpha_t(x) + c\alpha(x) = u_t(x) - \int_0^x r(x,y)u_t(y)\,dy - \int_0^x s(x,y)u_{xx}(y)\,dy\,, \tag{8.60}$$

$$\alpha_{xx}(x) = u_{xx}(x) + \int_0^x s(x,y)u_t(y)\,dy - \int_0^x r(x,y)u_{xx}(y)\,dy\,. \tag{8.61}$$

The controls are obtained by setting $x = 1$ in (8.60), (8.61):

$$u_t(1) = \int_0^1 r(1,y)u_t(y)\,dy + \int_0^1 s(1,y)u_{xx}(y)\,dy\,, \tag{8.62}$$

$$u_{xx}(1) = -\int_0^1 s(1,y)u_t(y)\,dy + \int_0^1 r(1,y)u_{xx}(y)\,dy\,. \tag{8.63}$$

Note that the feedback (8.62) would be implemented as integral, not proportional, control. Another observation we make is that even though the state α exponentially converges to zero, the same cannot be said about u. This is easy to see from (8.32): when v converges to zero, u may converge to an arbitrary constant.

Summarizing, we see that the direct application of the control design for the Schrödinger equation to the Euler–Bernoulli beam model results in control laws (8.62), (8.63) that guarantee the stabilization of the beam to a constant profile. This may be useful when the control objective is to suppress oscillations, without necessarily bringing the beam to the zero position.

8.2.4 Control Law That Guarantees Regulation to Zero

To achieve regulation to zero, we are going to modify the control law (8.62). Our objective is to express u_{xx} in (8.62) through the time derivatives u_t and u_{tt}.

We start by twice integrating the beam PDE (8.29) with respect to x, first from 0 to x, and then from x to 1. We get

$$u_{xx}(x) = u_{xx}(1) + \int_x^1 \int_0^y u_{tt}(\xi)\,d\xi\,dy\,. \tag{8.64}$$

Using the expression (8.63) for the control $u_{xx}(1)$, we get

$$u_{xx}(x) = -\int_0^1 s(1,y)u_t(y)\,dy + \int_0^1 r(1,y)u_{xx}(y)\,dy + \int_x^1 \int_0^y u_{tt}(\xi)\,d\xi\,dy\,. \tag{8.65}$$

Our objective is to express the second term in (8.65) through the terms with time derivatives. Let us multiply (8.65) by $r(1,x)$ and integrate from 0 to 1. We have

$$\begin{aligned}\int_0^1 r(1,y)u_{xx}(y)\,dy &= -\int_0^1 r(1,y)\,dy\left[\int_0^1 s(1,y)u_t(y)\,dy - \int_0^1 r(1,y)u_{xx}(y)\,dy\right] \\ &\quad + \int_0^1 r(1,y)\int_y^1 \int_0^z u_{tt}(\xi,t)\,d\xi\,dz\,dy\,. \end{aligned}\tag{8.66}$$

Therefore,

$$\begin{aligned}&\int_0^1 r(1,y)u_{xx}(y)\,dy \\ &\quad = \frac{\gamma_r - 1}{\gamma_r}\int_0^1 s(1,y)u_t(y)\,dy + \frac{1}{\gamma_r}\int_0^1 r(1,y)\int_y^1\int_0^z u_{tt}(\xi)\,d\xi\,dz\,dy \\ &\quad = \frac{\gamma_r - 1}{\gamma_r}\int_0^1 s(1,y)u_t(y)\,dy - \frac{1}{\gamma_r}\int_0^1 \left(R(1,y) - (1-\gamma_r)(1-y)\right)u_{tt}(y)\,dy, \end{aligned}\tag{8.67}$$

where we denote

$$R(x,y) = \int_y^x r(x,\xi)(\xi - y)\,d\xi\,, \tag{8.68}$$

$$\gamma_r = 1 - \int_0^1 r(1,y)\,dy = \cosh\left(\sqrt{\frac{c}{2}}\right)\cos\left(\sqrt{\frac{c}{2}}\right). \tag{8.69}$$

Substituting (8.67) into (8.65), after several simplifications and integrations by parts, we get

$$u_{xx}(x) = -\frac{1}{\gamma_r}\int_0^1 s(1,y)u_t(y)\,dy - \frac{1}{\gamma_r}\int_0^1 (R(1,y)-(1-\gamma_1)(1-y))\,u_{tt}(y)\,dy \\ + \int_x^1\int_0^y u_{tt}(\xi)\,d\xi\,dy\,. \tag{8.70}$$

Substituting (8.70) into (8.62) and noting that (after several integrations by parts)

$$\int_0^x s(x,y)\int_y^1\int_0^z u_{tt}(\xi)\,d\xi\,dz\,dy = -\gamma(x)\int_0^1 (1-y)u_{tt}(y)\,dy - \int_0^x S(x,y)u_{tt}(y)\,dy\,,$$

where

$$S(x,y) = \int_y^x s(x,\xi)(\xi - y)\,d\xi\,, \tag{8.71}$$

$$\gamma(x) = -\int_0^x s(x,y)\,dy = \sinh\left(\sqrt{\frac{c}{2}}x\right)\sin\left(\sqrt{\frac{c}{2}}x\right), \tag{8.72}$$

we finally obtain the following representation of (8.62):

$$u_t(1) = \int_0^1 \left(r(1,y) + \frac{\gamma(1)}{\gamma_r}s(1,y)\right)u_t(y)\,dy \\ - \int_0^1 \left[S(1,y) + \frac{\gamma(1)}{\gamma_r}(1-y-R(1,y))\right]u_{tt}(y)\,dy\,. \tag{8.73}$$

We now integrate (8.73) with respect to time to get the controller

$$u(1) = \int_0^1 \left(r(1,y) + \frac{\gamma(1)}{\gamma_r}s(1,y)\right)u(y)\,dy \\ - \int_0^1 \left[S(1,y) + \frac{\gamma(1)}{\gamma_r}(1-y-R(1,y))\right]u_t(y)\,dy\,. \tag{8.74}$$

The other controller (8.61) can also be represented in terms of u and u_t as follows:

$$u_{xx}(1) = -\int_0^1 s(1,y)u_t(y)\,dy + \frac{c^2}{8}u(1) + \int_0^1 r_{yy}(1,y)u(y)\,dy\,. \tag{8.75}$$

The control gains in (8.74) involve a division by γ_r, which may become zero for certain values of c. Therefore, c should satisfy the condition

$$c \neq \frac{\pi^2}{2}(2n+1)^2, \quad n = 0, 1, 2, \dots\,, \tag{8.76}$$

which is easily achieved because c is the designer's choice.

8.2.5 Explicit Form of Transformation

To find out what the actual transformation from u to α is, we start with the definition (8.52) and note that

$$\begin{aligned}\operatorname{Im}\{w(x)\} &= \operatorname{Im}\left\{v(x) - \int_0^x k(x,y)v(y)\,dy\right\}\\ &= -u_{xx}(x) + \int_0^x r(x,y)u_{xx}(y)\,dy - \int_0^x s(x,y)u_t(y)\,dy\,. \end{aligned} \tag{8.77}$$

We get

$$\begin{aligned}\alpha(x) = u(x) - u(1) &+ \int_x^1\int_0^y\int_0^z r(z,\xi)u_{xx}(\xi)\,d\xi dz dy\\ &- \int_x^1\int_0^y\int_0^z s(z,\xi)u_t(\xi)\,d\xi dz dy\,. \end{aligned} \tag{8.78}$$

Integrating by parts the term with u_{xx} and changing the order of integration in both integral terms, we obtain

$$\begin{aligned}\alpha(x) = u(x) &- \int_0^x \left(r(x,y) + c\bar{S}(x,y)\right)u(y)\,dy - \int_0^x \bar{S}(x,y)u_t(y)\,dy\\ &- u(1) + \int_0^1 \left(r(1,y) + c\bar{S}(1,y)\right)u(y)\,dy + \int_0^1 \bar{S}(1,y)u_t(y)\,dy\,, \end{aligned} \tag{8.79}$$

where

$$\bar{S}(x,y) = \int_y^x (x-\xi)s(\xi,y)\,d\xi\,. \tag{8.80}$$

Then, by substituting (8.74) into (8.79), we obtain the final form of the transformation:

$$\begin{aligned}\alpha(x) = u(x)&\\ &- \int_0^x \left(r(x,y) + c\bar{S}(x,y)\right)u(y)\,dy\\ &+ \int_0^1 \left(c\bar{S}(1,y) - \frac{\gamma(1)}{\gamma_r}s(1,y)\right)u(y)\,dy\\ &- \int_0^x \bar{S}(x,y)u_t(y)\,dy\\ &+ \int_0^1 \left[\bar{S}(1,y) + S(1,y) + \frac{\gamma(1)}{\gamma_r}(1 - y - R(1,y))\right]u_t(y)\,dy\,. \end{aligned} \tag{8.81}$$

Note that (8.81) is a novel type of transformation in this book—it is not of a strict-feedback form. It contains both spatially causal (Volterra) integrals and full-domain (Fredholm) integrals.

8.2.6 Convergence to a Zero Steady State

To show that (8.74), together with (8.75), stabilizes the plant to zero without computing the inverse transformation of (8.81), we first determine from (8.32) that u_t and u_{xx} converge to zero. Since $u_x(0) = 0$, this implies that u converges to a constant. Let us show that this constant is zero. Suppose $u(x,t) \equiv A$; then from (8.74) we get

$$0 = A\left(1 - \int_0^1 \left(r(1,y) + \frac{\gamma(1)}{\gamma_r}s(1,y)\right) dy\right). \tag{8.82}$$

Computing the integral on the right-hand side of (8.82), we obtain

$$0 = \frac{A}{\gamma_r}\left([\cosh(a)\cos(a)]^2 + [\sinh(a)\sin(a)]^2\right), \tag{8.83}$$

where $a = \sqrt{c/2}$. Using the identity $\sinh(a)^2 = \cosh(a)^2 - 1$, we can write (8.83) in the form

$$0 = \frac{A}{\gamma_r}\left(\cosh(a)^2 - \sin(a)^2\right). \tag{8.84}$$

Since $\cosh(a)^2 - \sin(a)^2 > 1$ for all $c > 0$, and $\gamma_r \neq 0$, we get $A = 0$.

8.2.7 Simulation Results

We present our simulation results by first summarizing the control law for the Euler–Bernoulli beam in Table 8.1. The gain functions $r(x,y)$ and $s(x,y)$ are shown in Figure 8.3. The results of numerical simulation are presented in Figures 8.4–8.6. The design parameter c was set to $c = 5$. In Figure 8.4 we can see the oscillations of the uncontrolled beam. With control, the beam is quickly brought to the zero equilibrium (Figure 8.5). The controls $u(1,t)$ and $u_{xx}(1,t)$ are shown in Figure 8.6. Note that the control effort is small (the maximum displacement of the controlled end is five times less than the initial displacement of the tip).

8.3 Notes and References

We highly recommend the survey on modeling of beams in [68]. As for control of beams, particularly those of Euler–Bernoulli type, comprehensive coverage of the topic is provided in the book of Luo, Guo, and Morgul [122].

The design for the shear beam in Section 8.1 was developed for the case of a free-end boundary condition at $x = 0$, but it can be extended to any other kind of a boundary condition (clamped, hinged, sliding) and even to the case of a destabilizing boundary condition (details are given in [96]).

The design for the Euler–Bernoulli beam in Section 8.2 is developed for a sliding-end boundary condition, $u_x(0) = u_{xxx}(0) = 0$. This design can also be extended to the case of a hinged boundary condition $u(0) = u_{xx}(0) = 0$. However, it is not clear at this point how to extend the design to the more basic cases of boundary conditions such as free-end

Table 8.1. *Control law for Euler–Bernoulli beam.*

Plant:

$$u_{tt} + u_{xxxx} = 0\,, \tag{8.85}$$

$$u_x(0) = 0\,, \tag{8.86}$$

$$u_{xxx}(0) = 0\,. \tag{8.87}$$

Controller:

$$\begin{aligned} u(1) &= \int_0^1 \left(r(1,y) + \frac{\gamma(1)}{\gamma_r} s(1,y) \right) u(y)\,dy\,, \\ &\quad - \int_0^1 \left[S(1,y) + \frac{\gamma(1)}{\gamma_r} (1 - y - R(1,y)) \right] u_t(y)\,dy\,, \end{aligned} \tag{8.88}$$

$$\begin{aligned} u_{xx}(1) &= \frac{c^2}{8} u(1) + \int_0^1 r_{yy}(1,y) u(y)\,dy\,, \\ &\quad - \int_0^1 s(1,y) u_t(y)\,dy\,. \end{aligned} \tag{8.89}$$

Gains:

$$r(x,y) = x\sqrt{\frac{c}{2\left(x^2-y^2\right)}} \left[-\mathrm{ber}_1\left(\sqrt{c\left(x^2-y^2\right)}\right) - \mathrm{bei}_1\left(\sqrt{c\left(x^2-y^2\right)}\right) \right], \tag{8.90}$$

$$s(x,y) = x\sqrt{\frac{c}{2\left(x^2-y^2\right)}} \left[\mathrm{ber}_1\left(\sqrt{c\left(x^2-y^2\right)}\right) - \mathrm{bei}_1\left(\sqrt{c\left(x^2-y^2\right)}\right) \right], \tag{8.91}$$

$$R(x,y) = \int_y^x r(x,\xi)(\xi - y)\,d\xi\,, \tag{8.92}$$

$$S(x,y) = \int_y^x s(x,\xi)(\xi - y)\,d\xi\,, \tag{8.93}$$

$$\gamma_r = \cosh\left(\sqrt{\frac{c}{2}}\right) \cos\left(\sqrt{\frac{c}{2}}\right), \tag{8.94}$$

$$\gamma(x) = \sinh\left(\sqrt{\frac{c}{2}}x\right) \sin\left(\sqrt{\frac{c}{2}}x\right). \tag{8.95}$$

and clamped. Nevertheless, it is possible to extend the design to a particular type of a destabilizing boundary condition at $x = 0$, as discussed in Exercise 8.3.

The control results in Figure 8.1 are obtained with observer-based feedback, where an observer state $\hat{u}(y,t)$, $\hat{u}_t(y,t)$ is used on the right-hand side of the control law (8.25).

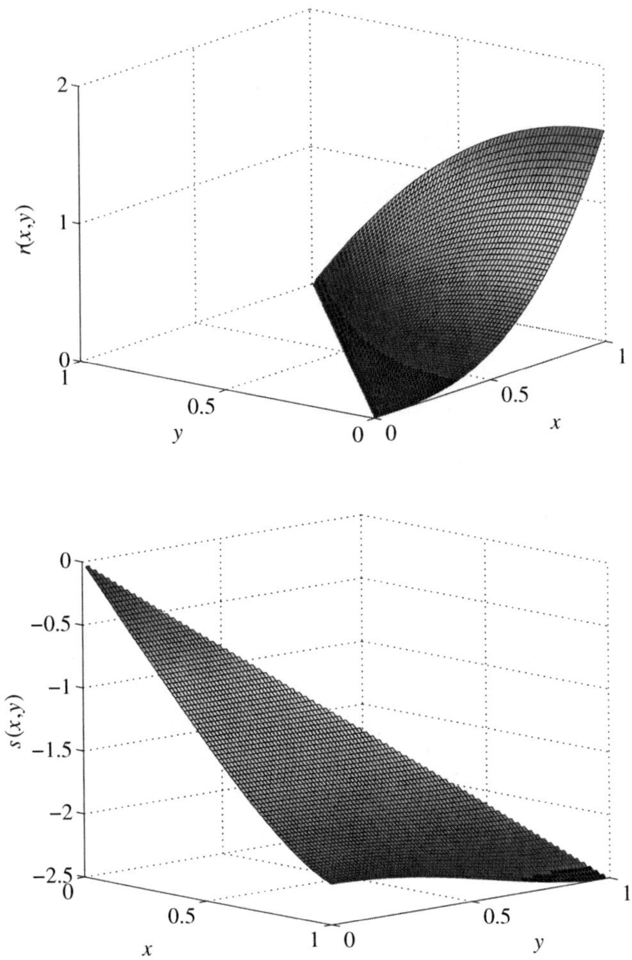

Figure 8.3. *The gain functions $r(x, y)$ and $s(x, y)$ for $c = 5$.*

We are not presenting the observer in this introductory book. Its design follows the ideas in Chapter 5 and is presented in [96].

The backstepping design can also be developed for the Timoshenko beam model. The Timoshenko model is given as

$$\varepsilon u_{tt} = (1 + d\partial_t)(u_{xx} - \theta_x)\,, \tag{8.96}$$

$$\mu\varepsilon\theta_{tt} = (1 + d\partial_t)(\varepsilon\theta_{xx} + a(u_x - \theta))\,, \tag{8.97}$$

where $u(x, t)$ denotes the displacement and $\theta(x, t)$ denotes the deflection angle. The positive constants a and μ are proportional to the nondimensional cross-sectional area, and the nondimensional moment of inertia of the beam, respectively. The parameter ε is inversely proportional to the nondimensional shear modulus of the beam. The coefficient d denotes the possible presence of Kelvin–Voigt damping. The Timoshenko beam can come with

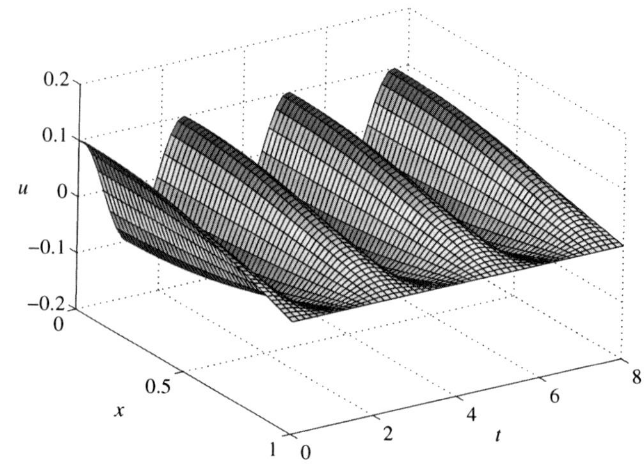

Figure 8.4. *Open-loop response of the Euler–Bernoulli beam.*

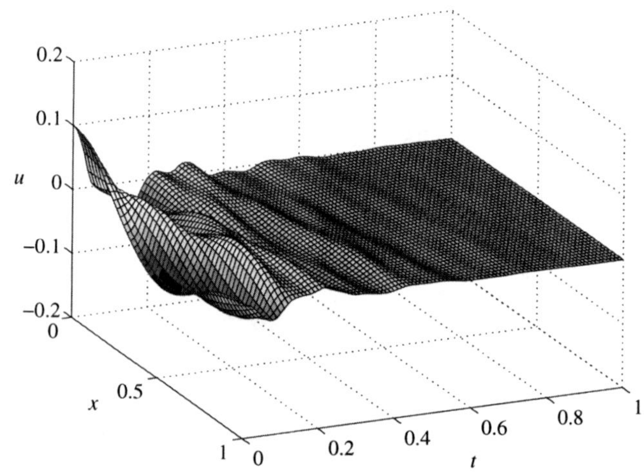

Figure 8.5. *Closed-loop response of the Euler–Bernoulli beam.*

any of the four basic types of boundary conditions. The free-end boundary condition is defined as

$$u_x(0,t) = \theta(0,t)\,, \tag{8.98}$$

$$\theta_x(0,t) = 0\,. \tag{8.99}$$

The meaning of the first equation is that zero force is being applied at the tip, whereas the meaning of the second equation is that zero moment is being applied at the tip. The relation between the Timoshenko beam and the shear beam is that the shear beam is a singular perturbation of the Timoshenko beam as $\mu \to 0$. Unlike the Euler–Bernoulli, shear, and Rayleigh beam models, which are second order in the time variable t, the Timoshenko beam

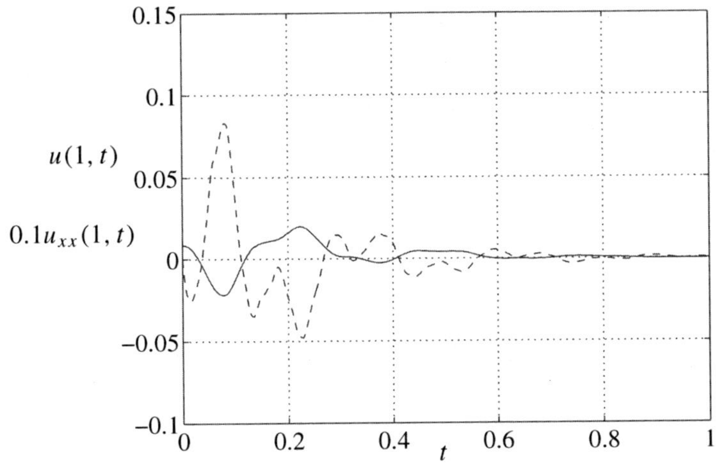

Figure 8.6. *Controls $u(1,t)$ (solid line) and $u_{xx}(1,t)$ (dashed line) for the Euler–Bernoulli beam.*

is fourth order in t. A backstepping boundary control design applied to the most complex beam model, the Timoshenko model, is presented in [99, 100].

Exercises

8.1. Derive the shear beam model

$$u_{tt} - \varepsilon u_{xxtt} + u_{xxxx} = 0$$

from the alternative model

$$\varepsilon u_{tt} = u_{xx} - \alpha_x \,,$$
$$0 = \varepsilon \alpha_{xx} - \alpha + u_x \,.$$

8.2. Show that

$$\alpha(x) = \cosh(bx)\alpha(0) - b\int_0^x \sinh(b(x-y))u_y(y)\,dy$$

is a solution of the equation

$$\alpha_{xx} - b^2\alpha + b^2 u_x = 0$$

with the initial condition $\alpha_x(0) = 0$.

8.3. Consider the Euler–Bernoulli beam model

$$u_{tt} + u_{xxxx} = 0$$

with a destabilizing boundary condition at the tip given by

$$u_{xx}(0) = -\frac{1}{q}u_{xt}(0),$$
$$u_{xxx}(0) = qu_t(0),$$

where $q > 0$. Show that the transformation

$$\alpha(x) = u(x) + (c_0 + q)\left[\frac{1-\cos(qx)}{q}u(0) - \frac{\sin(qx)}{q^2}u_x(0) - \int_0^x \sin(q(x-y))u(y)\,dy + \int_0^x \frac{\cos(q(x-y))-1}{q^2}u_t(y)\,dy\right]$$

along with the controls

$$u_{xx}(1) = (c_0 + q)\left[qu(1) - q\cos(q)u(0) - \sin(q)u_x(0) - q^2\int_0^1 \sin(q(1-y))u(y)\,dy + \int_0^1 \cos(q(1-y))u_t(y)\,dy\right],$$
$$u(1) = (c_0 + q)\left[\frac{\cos(q)-1}{q}u(0) + \frac{\sin(q)}{q^2}u_x(0) + \int_0^1 \sin(q(1-y))u(y)\,dy - \int_0^1 \frac{\cos(q(1-y))-1}{q^2}u_t(y)\,dy\right]$$

with $c_0 > 0$ converts the beam into the target system

$$\alpha_{tt} + \alpha_{xxxx} = 0,$$
$$\alpha_{xx}(0) = \frac{1}{c_0}\alpha_{xt}(0),$$
$$\alpha_{xxx}(0) = -c_0\alpha_t(0),$$
$$\alpha_{xx}(1) = 0,$$
$$\alpha(1) = 0.$$

Note that the gain kernels of the control laws are periodic functions of y, whose frequency grows with the increase of the "instability parameter" q.
Hint: Guided by the Schrödinger equation in Exercise 6.2, introduce

$$v = u_t - ju_{xx},$$
$$w = \alpha_t - j\alpha_{xx}$$

and consider the change of variable (6.72). Show that the transformation satisfies

$$\alpha_t(x) = u_t(x) - \int_0^x (c_0 + q)\sin(q(x-y))u_t(y)\,dy + \int_0^x (c_0 + q)\cos(q(x-y))u_{xx}(y)\,dy\,, \tag{8.100}$$

$$\alpha_{xx}(x) = u_{xx}(x) - \int_0^x (c_0 + q)\cos(q(x-y))u_t(y)\,dy - \int_0^x (c_0 + q)\sin(q(x-y))u_{xx}(y)\,dy\,. \tag{8.101}$$

Integrate by parts the second integral in (8.100) to get u_{xxxx} under the integral sign. Use the plant and the boundary conditions to express all the terms on the right-hand side of (8.100) through the terms with time derivatives. Integrate the result with respect to time.

8.4. Consider the plant from Exercise 8.3 for $q = 0$:

$$u_{tt} + u_{xxxx} = 0\,, \tag{8.102}$$

$$u_{xt}(0) = 0\,, \tag{8.103}$$

$$u_{xxx}(0) = 0\,. \tag{8.104}$$

(a) Show that the transformation

$$\alpha(x) = u(x) - \frac{c_0}{2}\int_0^x (x-y)^2 u_t(y)\,dy + c_0 x \int_0^1 (1-y)u_t(y)\,dy + c_0^2 x \int_0^1 u(y)\,dy \tag{8.105}$$

along with the controls

$$u(1) = -\frac{c_0}{2}\int_0^1 (1-y^2)u_t(y)\,dy - c_0^2 \int_0^1 u(y)\,dy \tag{8.106}$$

$$u_{xx}(1) = c_0 \int_0^1 u_t(y)\,dy \tag{8.107}$$

maps (8.102)–(8.104) into the target system

$$\alpha_{tt} + \alpha_{xxxx} = 0\,,$$

$$\alpha_{xx}(0) = \frac{1}{c_0}\alpha_{xt}(0)\,,$$

$$\alpha_{xxx}(0) = -c_0\alpha_t(0)\,,$$

$$\alpha_{xx}(1) = 0\,,$$

$$\alpha(1) = 0\,.$$

Hint: Start by showing that (8.105), together with (8.106) and (8.107), gives

$$\alpha_t(x) = u_t(x) + c_0 \int_0^x u_{xx}(y)\,dy\,, \tag{8.108}$$

$$\alpha_{xx}(x) = u_{xx}(x) - c_0 \int_0^x u_t(y)\,dy\,. \tag{8.109}$$

(b) Show that the inverse transformation

$$u(x) = \alpha(x) - c_0 \int_0^x \sin(c_0(x-y))\alpha(y)\,dy - (c_0 \sin(c_0 x) + \cos(c_0 x) - 1)\,\alpha(0)$$
$$+\frac{1}{c_0}\int_0^x [1-\cos(c_0(x-y))]\alpha_t(y)\,dy - \sin(c_0 x)\int_0^1 (1-y)\alpha_t(y)\,dy$$

satisfies

$$\begin{aligned} u_t(x) &= \alpha_t(x) - c_0 \int_0^x \sin(c_0(x-y))\alpha_t(y)\,dy \\ &\quad - c_0 \int_0^x \cos(c_0(x-y))\alpha_{xx}(y)\,dy\,, \\ u_{xx}(x) &= \alpha_{xx}(x) + c_0 \int_0^x \cos(c_0(x-y))\alpha_t(y)\,dy \\ &\quad - c_0 \int_0^x \sin(c_0(c-y))\alpha_{xx}(y)\,dy\,. \end{aligned}$$

Chapter 9

First-Order Hyperbolic PDEs and Delay Equations

In the previous two chapters we considered hyperbolic PDEs of second order that usually describe oscillatory systems such as strings and beams. We now turn our attention to the first-order hyperbolic equations. They describe quite a different set of physical problems, such as chemical reactors, heat exchangers, and traffic flows. They can also serve as models for delays.

9.1 First-Order Hyperbolic PDEs

The general first-order hyperbolic equation that can be handled by the backstepping method is

$$u_t = u_x + g(x)u(0) + \int_0^x f(x,y)u(y)dy\,, \tag{9.1}$$

$$u(1) = \text{control}\,. \tag{9.2}$$

Unlike in second-order PDEs, here we specify only one boundary condition. For g or f positive and large, this system (with zero boundary condition at $x = 1$) can become unstable.

Following our general procedure, we use the transformation

$$w(x) = u(x) - \int_0^x k(x,y)u(y)dy \tag{9.3}$$

to convert (9.1) into the target system

$$w_t = w_x\,, \tag{9.4}$$

$$w(1) = 0\,. \tag{9.5}$$

This system is a delay line with unit delay, output $w(0,t) = w(1,t-1)$, and zero input at $w(1,t)$, akin to traffic flow over a stretch of road on which no additional cars are permitted to enter after $t = 0$. Its solution is

$$w(x,t) = \begin{cases} w_0(t+x) & 0 \le t+x < 1\,, \\ 0 & t+x \ge 1\,, \end{cases} \tag{9.6}$$

where $w_0(x)$ is the initial condition. We see that this solution becomes zero in finite time.

One can derive the following kernel PDE from (9.1)–(9.5):

$$k_x(x,y) + k_y(x,y) = \int_y^x k(x,\xi) f(\xi,y) d\xi - f(x,y)\,, \tag{9.7}$$

$$k(x,0) = \int_0^x k(x,y) g(y) dy - g(x)\,. \tag{9.8}$$

This is a well posed equation, and after its solution is found, the controller, as before, is given by

$$u(1) = \int_0^1 k(1,y) u(y)\, dy\,. \tag{9.9}$$

Example 9.1 Consider the plant

$$u_t = u_x + g e^{bx} u(0)\,, \tag{9.10}$$

where g and b are constants. In this case, equation (9.7) becomes

$$k_x + k_y = 0\,, \tag{9.11}$$

which has a general solution $k(x,y) = \phi(x-y)$. If we plug this solution into (9.8), we get the integral equation

$$\phi(x) = \int_0^x g e^{by} \phi(x-y) dy - g e^{bx}\,. \tag{9.12}$$

The solution to this equation can be easily obtained by applying the Laplace transform in x:

$$\phi(s) = -\frac{g}{s-b-g}\,, \tag{9.13}$$

and after taking the inverse Laplace transform, $\phi(x) = -g e^{(b+g)x}$. Therefore, the solution to the kernel PDE is $k(x,y) = -g e^{(b+g)(x-y)}$ and the controller is given by (9.9). $\diamond$

Example 9.2 Consider the plant

$$u_t = u_x + \int_0^x f e^{b(x-y)} u(y)\, dy\,, \tag{9.14}$$

where f and b are constants. The kernel PDE takes the form

$$k_x + k_y = \int_y^x k(x,\xi) f e^{b(\xi-y)} d\xi - f e^{b(x-y)}\,, \tag{9.15}$$

$$k(x,0) = 0\,. \tag{9.16}$$

After we differentiate (9.15) with respect to y, the integral term is eliminated:

$$k_{xy} + k_{yy} = -fk - bk_x - bk_y\,. \tag{9.17}$$

Since we now increased the order of the equation, we need an extra boundary condition. We get it by setting $y = x$ in (9.15):

$$\frac{d}{dx}k(x, x) = k_x(x, x) + k_y(x, x) = -f, \tag{9.18}$$

which, after integration, becomes $k(x, x) = -fx$.

Introducing the change of variables

$$k(x, y) = p(z, y)e^{b(z-y)/2}, \qquad z = 2x - y\,, \tag{9.19}$$

we get the following PDE for $p(z, y)$:

$$p_{zz}(z, y) - p_{yy}(z, y) = fp(z, y)\,, \tag{9.20}$$

$$p(z, 0) = 0\,, \tag{9.21}$$

$$p(z, z) = -fz\,. \tag{9.22}$$

This PDE has been solved in Chapter 4. We get

$$p(z, y) = -2fy\frac{I_1\left(\sqrt{f(z^2 - y^2)}\right)}{\sqrt{f(z^2 - y^2)}} \tag{9.23}$$

or, in the original variables,

$$k(x, y) = -fe^{b(x-y)}y\frac{I_1\left(2\sqrt{fx(x - y)}\right)}{\sqrt{fx(x - y)}}\,, \tag{9.24}$$

and the Dirichlet controller is given by (9.9). ◇

9.2 ODE Systems with Actuator Delay

Consider a linear finite-dimensional system with the actuator delay

$$\dot{X} = AX + BU(t - D)\,, \tag{9.25}$$

where (A, B) is a controllable pair and the input signal $U(t)$ is delayed by D units of time. Because of the delay, the problem is infinite-dimensional, and this is where PDEs come into play. The following first-order hyperbolic PDE serves as a model for the delay:

$$u_t = u_x\,, \tag{9.26}$$

$$u(D, t) = U(t)\,. \tag{9.27}$$

The solution to this equation is $u(x, t) = U(t + x - D)$, and therefore the output $u(0, t) = U(t - D)$ gives the delayed input. The system (9.25) can be now written as

$$\dot{X} = AX + Bu(0, t). \tag{9.28}$$

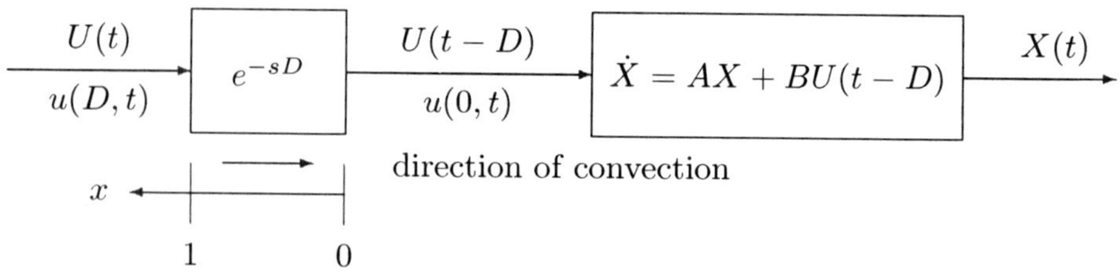

Figure 9.1. *Linear system $\dot{X} = AX + BU(t-D)$ with actuator delay D.*

The equations (9.26)–(9.28) form an ODE–PDE cascade which is driven by the input U from the boundary of the PDE (Figure 9.1).

Suppose a static state-feedback control has been designed for a system with no delay (i.e., with $D = 0$) such that $U = KX$ is a stabilizing controller, i.e., the matrix $(A + BK)$ is Hurwitz. When $D \neq 0$, we choose the target system as follows:

$$\dot{X} = (A + BK)X + Bw(0)\,, \tag{9.29}$$

$$w_t = w_x\,, \tag{9.30}$$

$$w(D) = 0\,. \tag{9.31}$$

This choice of the target system is natural: w becomes zero in finite time, and after that the ODE (9.29) is guaranteed to be exponentially stable by the nominal design.

To map (9.26)–(9.28) into (9.29)–(9.31), consider the backstepping transformation

$$w(x) = u(x) - \int_0^x q(x,y)u(y)dy - \gamma(x)^T X\,, \tag{9.32}$$

where $q(x, y)$ and $\gamma(x)$ are to be designed. As usual, let us calculate the time and spatial derivatives of the transformation (9.32):

$$w_x = u_x - q(x,x)u(x) - \int_0^x q_x(x,y)u(y)dy - \gamma'(x)^T X\,, \tag{9.33}$$

$$\begin{aligned} w_t &= u_t - \int_0^x q(x,y)u_t(y)dy - \gamma(x)^T\left[AX + Bu(0)\right] \\ &= u_x - q(x,x)u(x) + q(x,0)u(0) + \int_0^x q_y(x,y)u(y)dy \\ &\quad -\gamma(x)^T\left[AX + Bu(0)\right]\,. \end{aligned} \tag{9.34}$$

Subtracting (9.33) from (9.34), we get

$$\begin{aligned} \int_0^x (q_x(x,y) + q_y(x,y))u(y)dy + \left[q(x,0) - \gamma(x)^T B\right]u(0) & \\ + \left[\gamma'(x)^T - \gamma(x)^T A\right]X = 0\,. & \end{aligned} \tag{9.35}$$

This equation must be verified for all u and X, so we have three conditions:

$$q_x(x,y) + q_y(x,y) = 0\,, \tag{9.36}$$
$$q(x,0) = \gamma(x)^T B\,, \tag{9.37}$$
$$\gamma'(x) = A^T \gamma(x)\,. \tag{9.38}$$

The first two conditions form a familiar first-order hyperbolic PDE, and the third one is an ODE. To find the initial condition for this ODE, let us set $x = 0$ in (9.32), which gives $w(0) = u(0) - \gamma(0)^T X$. Substituting this expression into (9.29), we get

$$\dot{X} = AX + Bu(0) + B\left(K - \gamma(0)^T\right)X\,. \tag{9.39}$$

Comparing this equation with (9.28), we have $\gamma(0) = K^T$. Therefore the solution to the ODE (9.38) is $\gamma(x) = e^{A^T x}K^T$, which gives

$$\gamma(x)^T = Ke^{Ax}\,. \tag{9.40}$$

A general solution to (9.36) is $q(x,y) = \phi(x-y)$, where the function ϕ is determined from (9.37). We get

$$q(x,y) = Ke^{A(x-y)}B\,. \tag{9.41}$$

We can now substitute the gains $\gamma(x)$ and $q(x,y)$ into the transformation (9.32) and set $x = D$ to get the control law:

$$\boxed{u(D) = K\int_0^D e^{A(D-y)}Bu(y)dy + Ke^{AD}X\,.} \tag{9.42}$$

The above controller is given in terms of the transport delay state $u(y)$. Using (9.26)–(9.27), one can also derive the representation in terms of the input signal $U(t)$:

$$\boxed{U(t) = K\left[e^{AD}X + \int_{t-D}^{t} e^{A(t-\theta)}BU(\theta)d\theta\right].} \tag{9.43}$$

Notice that this is an infinite-dimensional controller.

Using the same approach as in Chapter 5, one can also design observers for ODEs with sensor delay; see Exercises 9.2 and 9.3.

9.3 Notes and References

The controller (9.43) is the analog of the Smith Predictor (1957) extended to unstable plants and was first derived in 1978–1982 [102], [126], [6]. The derivation in these references is very different and employs a transformation of the ODE state rather than the delay state. As a result, the analysis in these references establishes convergence of the state to zero or eigenspectrum properties of the closed-loop system but does not actually capture *stability* of the entire ODE + delay system (in the sense of the effect of initial conditions, decay rate, and overshoot) as does (9.29)–(9.31). The backstepping procedure outlined in this chapter is more general and is applicable, for example, to parabolic PDEs with actuator and sensor delays, which is a subject of current research.

Exercises

9.1. Derive the kernel PDE (9.7)–(9.8).
Hint: use the formula

$$\int_0^x \int_0^\xi k(x,\xi) f(\xi, y) u(y)\, dy\, d\xi = \int_0^x \int_y^x k(x,\xi) f(\xi, y) u(y)\, d\xi\, dy\,.$$

9.2. Consider the system

$$\dot{X} = AX\,,$$
$$Y(t) = CX(t-D)\,,$$

where the output equation can be also represented as

$$u_t = u_x\,,$$
$$u(D,t) = CX(t)\,,$$
$$Y(t) = u(0,t)\,.$$

Introduce the observer

$$\dot{\hat{X}} = A\hat{X} + e^{AD} L\left(Y(t) - \hat{u}(0,t)\right)\,,$$
$$\hat{u}_t = \hat{u}_x + Ce^{Ax} L\left(Y(t) - \hat{u}(0,t)\right)\,,$$
$$\hat{u}(D,t) = C\hat{X}(t)\,,$$

where L is chosen such that $A - LC$ is Hurwitz. Show that the transformation

$$\tilde{w}(x) = \tilde{u}(x) - Ce^{A(x-D)}\tilde{X}\,,$$

where $\tilde{X} = X - \hat{X}$, $\tilde{u} = u - \hat{u}$, converts the $(\tilde{X}, \tilde{u})$ system into

$$\dot{\tilde{X}} = \left(A - e^{AD} L C e^{-AD}\right)\tilde{X} - e^{AD} L \tilde{w}(0)\,,$$
$$\tilde{w}_t = \tilde{w}_x\,,$$
$$\tilde{w}(D) = 0\,.$$

Note that the $\tilde{w}$ system is exponentially stable and that the matrix $A - e^{AD}LCe^{-AD}$ is Hurwitz (you can see this by using a similarity transformation e^{AD} and using the fact that it commutes with A).

9.3. Show that the observer in Exercise 9.2 can be represented as

$$\dot{\hat{X}} = A\hat{X} + e^{AD} L(Y - \hat{Y})\,,$$
$$\hat{Y}(t) = C\hat{X}(t-D) + C\int_{t-D}^{t} e^{A(t-\theta)} L(Y(\theta) - \hat{Y}(\theta))\, d\theta\,.$$

Hint: Take a Laplace transform of the $\hat{u}(x,t)$ system with respect to t; solve the resulting first-order ODE w.r.t. x with $\hat{u}(0,s) = \hat{Y}(s)$ as initial condition and $Y(s) - \hat{Y}(s)$ as input; evaluate the solution at $x = D$ and substitute $\hat{u}(D,s) = C\hat{X}(s)$; take the inverse Laplace transform; obtain the delayed versions of $\hat{X}(t)$ and $Y(t) - \hat{Y}(t)$; shift the integration variable to obtain $\int_{t-D}^{t}$.

Chapter 10

Kuramoto–Sivashinsky, Korteweg–de Vries, and Other "Exotic" Equations

Two equations that are popular in the research communities studying chaos, strange attractors in PDEs, and soliton waves are the Kuramoto–Sivashinsky and Korteweg–de Vries equations. The former is frequently used as a model of flamefront instabilities and thin film growth instabilities, whereas the latter is used as a model of shallow water waves and ion-acoustic waves in plasmas.

The Kuramoto–Sivashinsky equation is

$$\omega_t + \delta\omega_{xxxx} + \lambda\omega_{xx} + \omega_x\omega = 0\,, \tag{10.1}$$

whereas the Korteweg–de Vries equation is

$$\omega_t + \delta\omega_{xxx} + \lambda\omega_x + \omega_x\omega = 0\,, \tag{10.2}$$

where δ is positive and λ can be either sign. Even after we linearize these two equations by dropping the quadratic convective terms $\omega_x\omega$, these equations represent challenging control problems. In addition to linearizing them, we are going to add a new effect/term to these equations, which will make them easily tractable using the backstepping ideas introduced in Chapters 4, 8, and 9. We will consider a "Kuramoto–Sivashinsky-like" linear PDE

$$u_t - \nu u_{txx} + \delta u_{xxxx} + \lambda u_{xx} = 0\,, \tag{10.3}$$

and a "Korteweg–de Vries-like" linear PDE

$$u_t - \nu u_{txx} + \delta u_{xxx} + \lambda u_x = 0\,, \tag{10.4}$$

where $\nu > 0$. When ν is small, these systems behave almost the same as the respective linearized Kuramoto–Sivashinsky and Korteweg–de Vries equations and can be unstable when λ/δ is positive and large. Let us denote the following new constants:

$$a = \frac{1}{\delta}, \quad \varepsilon = \frac{\nu}{\delta}, \quad \gamma = 1 + \varepsilon\lambda\,. \tag{10.5}$$

Then we can write (10.3) as

$$\varepsilon u_t = u_{xx} - v_x\,, \tag{10.6}$$

$$0 = \varepsilon v_{xx} + a(-v + \gamma u_x) \tag{10.7}$$

and (10.4) as

$$\varepsilon u_t = u_x - v\,, \tag{10.8}$$
$$0 = \varepsilon v_{xx} + a(-v + \gamma u_x)\,. \tag{10.9}$$

One should note a striking similarity between these equations and the shear beam model (8.9), (8.10). Furthermore, for simplicity we adopt the same boundary conditions (8.11), (8.12) for (10.6), (10.7), namely,

$$u_x(0) = v(0)\,, \tag{10.10}$$
$$v_x(0) = 0\,, \tag{10.11}$$

whereas for system (10.8), (10.9), which is one order lower, we need one less boundary condition at $x = 0$, and thus we adopt

$$v_x(0) = 0\,. \tag{10.12}$$

In the next two sections we present control designs for the above systems (10.6), (10.7), (10.10), (10.11) and (10.8), (10.9), (10.12), assuming two Dirichlet inputs $v(1)$, $u(1)$ for both systems.

10.1 Kuramoto–Sivashinsky Equation

The solution to (10.7), (10.11) is given by

$$v(x) = \cosh(bx)v(0) - \gamma b \int_0^x \sinh(b(x-y))u_y(y)\,dy\,, \tag{10.13}$$

where

$$b = \sqrt{\frac{a}{\varepsilon}}\,. \tag{10.14}$$

The first control

$$v(1) = \gamma b\left(\sinh(b)u(0) - b\int_0^1 \cosh(b(1-y))u(y)\,dy\right) \tag{10.15}$$

guarantees that

$$v(0) = 0 \tag{10.16}$$

and transforms (10.6), (10.7), (10.10) into the form

$$\varepsilon u_t = u_{xx} + \gamma\left(b^2 u - b^2\cosh(bx)u(0) + b^3\int_0^x \sinh(b(x-y))u(y)\,dy\right), \tag{10.17}$$
$$u_x(0) = 0\,. \tag{10.18}$$

Taking a feedback transformation

$$w(x) = u(x) - \int_0^x k(x,y)u(y)\,dy \tag{10.19}$$

and second control

$$u(1) = \int_0^1 k(1, y)u(y)\,dy \tag{10.20}$$

with a $k(x, y)$ that satisfies

$$k_{xx} - k_{yy} = \gamma b^2 \left(k + b\sinh(b(x-y)) - b\int_y^x k(x,\xi)\sinh(b(\xi - y))\,d\xi \right), \tag{10.21}$$

$$k_y(x,0) = \gamma b^2 \left(\int_0^x k(x,y)\cosh(by)\,dy - \cosh(bx) \right), \tag{10.22}$$

$$k(x,x) = -\frac{\gamma b^2}{2}x, \tag{10.23}$$

we convert (10.17), (10.18), (10.20) into the exponentially stable heat equation

$$\varepsilon w_t = w_{xx}, \tag{10.24}$$

$$w_x(0) = 0, \tag{10.25}$$

$$w(1) = 0. \tag{10.26}$$

The entire calculation (10.17)–(10.26) is a straightforward application of the backstepping design in Section 4.9.

10.2 Korteweg–de Vries Equation

Since (10.7) and (10.9) are the same and (10.11) and (10.12) are the same, we start with the transformation (10.13) and first control (10.15), which transform the system (10.8), (10.9) into the system

$$\varepsilon u_t = u_x - \gamma b \left(\sinh(bx)u(0) - b\int_0^x \cosh(b(x-y))u(y)\,dy \right). \tag{10.27}$$

The feedback transformation (10.19), (10.20) with $k(x, y)$ satisfying

$$k_x + k_y = -\gamma b^2 \left(\int_y^x k(x,\xi)\cosh(b(\xi - y))\,d\xi + \cosh(b(x-y)) \right), \tag{10.28}$$

$$k(x,0) = \gamma b \left(-\int_0^x k(x,y)\sinh(by)\,dy + \sinh(bx) \right) \tag{10.29}$$

converts (10.27), (10.20) into the exponentially stable transport equation

$$\varepsilon w_t = w_x, \tag{10.30}$$

$$w(1) = 0. \tag{10.31}$$

This result is a direct consequence of the design for first-order hyperbolic equations in Section 9.1.

10.3 Notes and References

Kuramoto–Sivashinsky and Korteweg–de Vries equations are rich topics for control research. Nonlinear boundary control of these equations (under some parametric restrictions) was considered in [115, 117, 10], where an abundance of other references is also provided, particularly on the analysis of these nonlinear PDEs.

Besides being related to the Korteweg–de Vries PDE, equation (10.4) also can be obtained as an approximation of the linearized Boussinesq PDE system modeling complex water waves such as tidal bores [57].

Exercises

10.1. Derive (10.3) from (10.6), (10.7).

10.2. Derive (10.4) from (10.8), (10.9).

Chapter 11

Navier–Stokes Equations

Navier–Stokes equations, which model turbulent fluid flows, are among the most exciting problems in applied mathematics and arguably are the most challenging problem in classical mechanics. Control of turbulent fluid flows is an active area of current research, with many problems being considered and many methods being developed. In this chapter we will illustrate the methods we have developed in the previous chapters by applying them to linearized Navier–Stokes equations in a 3D "channel flow" geometry.

11.1 Channel Flow PDEs and Their Linearization

We consider incompressible, nondimensionalized Navier–Stokes equations:

$$\mathbf{u}_t = \frac{1}{Re}\nabla^2\mathbf{u} - \mathbf{u}\cdot\nabla\mathbf{u} - \nabla p\,, \tag{11.1}$$

$$\nabla\cdot\mathbf{u} = 0\,, \tag{11.2}$$

where $\mathbf{u}$ is the velocity vector, p is the pressure, Re is the Reynolds number, and ∇ denotes the gradient. Equation (11.1) shows that, as the velocities develop, they are affected not only by a diffusion term $\nabla^2\mathbf{u}$, an advective term $\mathbf{u}\cdot\nabla\mathbf{u}$ but also by the pressure gradient ∇p. Equation (11.2) imposes an algebraic constraint on the system and comes from the continuity equation and the incompressible flow constraint.

We shall study the flow in a channel that is infinite in the x and z directions, is bounded at $y = 0$ and $y = 1$ by walls, and is forced by a pressure gradient in the x direction (Figure 11.1). This kind of flow is called the Poiseuille flow. The equilibrium solution to the Poiseuille flow is

$$U^e = 4y(1-y)\,, \tag{11.3}$$

$$V = W = 0\,, \tag{11.4}$$

$$P^e = P_0 - \frac{8}{Re}x\,, \tag{11.5}$$

where U, V, and W are the flows in the x, y, and z directions, respectively.

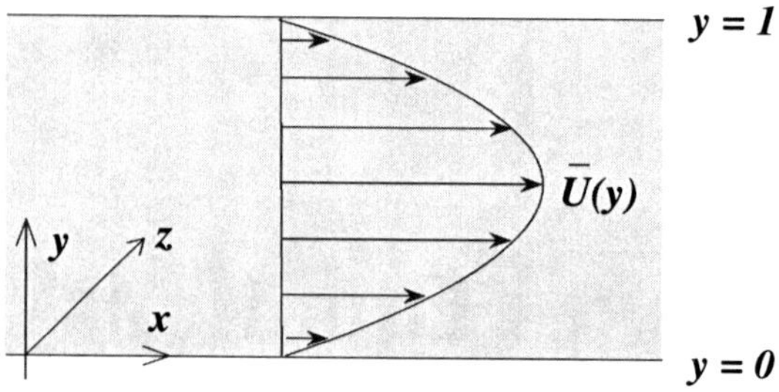

Figure 11.1. *Poiseuille flow in a 3D channel.*

There are two conditions which govern the behavior at the walls. These are the no-slip and no-penetration conditions. The no-slip condition ensures that the flow in the tangential directions (x and z directions) is zero at the walls. As the wall is solid, we cannot have fluid penetrating into the wall; therefore the flow in the normal direction (the y direction) must be zero at the walls also. We therefore have

$$\mathbf{u}(y=0) = 0\,, \tag{11.6}$$

$$\mathbf{u}(y=1) = 0 \tag{11.7}$$

for boundary conditions.

Is this flow stable? The Reynolds number Re determines stability. A low Re corresponds to a laminar (stable) flow, whereas a high Re corresponds to a turbulent (unstable) flow which does not remain at the equilibrium (11.3)–(11.5). Physically, Re is the ratio of the inertial forces to the viscous forces, proportional to the mean velocity through the channel and the width of the channel, and inversely proportional to the kinematic viscosity. We would like to stabilize the system around the Poiseuille equilibrium using control inputs distributed along the top wall ($y = 1$); the control inputs are implemented by microjet actuators for blowing and suction.

The first step in this analysis is to introduce the perturbation variables

$$u = U - U^e\,,$$
$$V = V - 0\,,$$
$$W = W - 0\,,$$
$$p = P - P^e\,,$$

which are governed by

$$u_t = \frac{1}{Re}\nabla^2 u - (u + U^e)u_x - V(u_y + U^e_y) - Wu_z - p_x\,, \tag{11.8}$$

$$V_t = \frac{1}{Re}\nabla^2 V - (u + U^e)V_x - VV_y - WV_z - p_y\,, \tag{11.9}$$

$$W_t = \frac{1}{Re}\nabla^2 W - (u + U^e)W_x - VW_y - WW_z - p_z\,, \tag{11.10}$$

$$u_x + W_z + V_y = 0 \tag{11.11}$$

with boundary conditions at the uncontrolled wall given by

$$u|_{y=0)} = V|_{y=0} = W|_{y=0} = 0 \tag{11.12}$$

and the boundary conditions at the controlled wall given by

$$u|_{y=1} = U_c(t,x,z)\,, \tag{11.13}$$
$$V|_{y=1} = V_c(t,x,z)\,, \tag{11.14}$$
$$W|_{y=1} = W_c(t,x,z)\,, \tag{11.15}$$

where U_c, V_c, W_c are our control inputs. After linearizing the perturbation system, the equations become

$$u_t = \frac{1}{Re}(u_{xx} + u_{zz} + u_{yy}) - U^e u_x - U^e_y V - p_x\,, \tag{11.16}$$
$$V_t = \frac{1}{Re}(V_{xx} + V_{zz} + V_{yy}) - U^e V_x - p_y\,, \tag{11.17}$$
$$W_t = \frac{1}{Re}(W_{xx} + W_{zz} + W_{yy}) - U^e W_x - p_z\,, \tag{11.18}$$
$$u_x + W_z + V_y = 0 \tag{11.19}$$

with (11.12)–(11.15) still holding for boundary conditions.

11.2 From Physical Space to Wavenumber Space

The next step is to transform the equations to the wave space. We take the 2D Fourier transform

$$\tilde{F}(t,k_x,y,k_z) = \int_{-\infty}^{\infty}\int_{-\infty}^{\infty} F(t,x,y,z)\mathrm{e}^{-2\pi j(k_x x + k_z z)}dxdz \tag{11.20}$$

in the x and z directions, as these directions are infinite. We started with a nonlinear 3D system, which we turned into a linear 3D system. After taking the Fourier transform, we will have an infinite number of uncoupled 1D systems parameterized by k_x and k_z, which are the wavenumbers in the x and z directions, respectively. One can use this approach for other applications as well, such as infinite or periodic thin plates.

Recalling the useful property of the Fourier transform under differentiation,

$$F_x = 2\pi j k_x \tilde{F}\,, \tag{11.21}$$
$$F_z = 2\pi j k_z \tilde{F}\,, \tag{11.22}$$

we see that the transform of (11.16)–(11.19) is

$$u_t = \frac{1}{Re}(-\alpha^2 u + u_{yy}) + \frac{\beta}{2}y(y-1)u + 4(2y-1)V - 2\pi j k_x p\,, \tag{11.23}$$
$$W_t = \frac{1}{Re}(-\alpha^2 W + W_{yy}) + \frac{\beta}{2}y(y-1)W - 2\pi j k_z p\,, \tag{11.24}$$
$$V_t = \frac{1}{Re}(-\alpha^2 V + V_{yy}) + \frac{\beta}{2}y(y-1)V - p_y\,, \tag{11.25}$$
$$2\pi j k_x u + 2\pi j k_z W + V_y = 0\,, \tag{11.26}$$

where

$$\alpha^2 = 4\pi^2(k_x^2 + k_z^2)\,, \tag{11.27}$$

$$\beta = 16\pi j k_x\,. \tag{11.28}$$

The transformed boundary conditions are

$$u|_{y=0)} = V|_{y=0} = W|_{y=0} = 0\,, \tag{11.29}$$

$$u|_{y=1} = U_c(t, k_x, k_z)\,, \tag{11.30}$$

$$V|_{y=1} = V_c(t, k_x, k_z)\,, \tag{11.31}$$

$$W|_{y=1} = W_c(t, k_x, k_z)\,. \tag{11.32}$$

For notational clarity we drop the dependence on k_x and k_z: $u(t, y) = u(t, k_x, y, k_z)$.

11.3 Control Design for Orr–Sommerfeld and Squire Subsystems

To continue the derivation, we employ a standard transformation from flow stability analysis, and write the flow equations in terms of normal velocity and vorticity, with one caveat: We consider V_y instead of V. Recalling (11.26), we define the new variables Y and ω as

$$Y = V_y \;=\; -2\pi j(k_x u + k_z W)\,, \tag{11.33}$$

$$\omega = -2\pi j(k_z u - k_x W)\,. \tag{11.34}$$

The original variables are expressed through Y and ω as follows:

$$u = \frac{-1}{2\pi j}\frac{k_x Y + k_z\omega}{k_x^2 + k_z^2}\,, \tag{11.35}$$

$$W = \frac{-1}{2\pi j}\frac{k_z Y - k_x\omega}{k_x^2 + k_z^2}\,, \tag{11.36}$$

$$V = \int_0^y Y(\eta)d\eta\,. \tag{11.37}$$

We can see from (11.35)–(11.37) that, to regulate the original variables to zero, it is sufficient to regulate Y and ω to zero.

The equations for Y and ω (the so-called Orr–Sommerfeld and Squire subsystems) are derived from (11.23) and (11.24):

$$Y_t = \frac{1}{Re}(-\alpha^2 Y + Y_{yy}) + \frac{\beta}{2}y(y-1)Y - 8\pi j k_x(2y-1)V - \alpha^2 p\,, \tag{11.38}$$

$$\omega_t = \frac{1}{Re}(-\alpha^2\omega + \omega_{yy}) + \frac{\beta}{2}y(y-1)\omega + 8\pi j k_z(2y-1)V \tag{11.39}$$

with boundary conditions

$$Y|_{y=0} = \omega|_{y=0} = 0\,, \tag{11.40}$$

$$Y|_{y=1} = -2\pi j(k_x U_c + k_z W_c)\,, \tag{11.41}$$

$$\omega|_{y=1} = -2\pi j(k_z U_c - k_x W_c)\,. \tag{11.42}$$

To employ the backstepping method, the system needs to be autonomous, i.e., Y_t and ω_t must be in terms of only Y and ω. The pressure term, p, can be replaced by a function of V by deriving an equation for p and solving this equation in terms of V. We take the divergence of the velocity equations (11.23)–(11.24), apply (11.26), and arrive at the following equation for the pressure:

$$-\alpha^2 p + p_{yy} = \beta(2y-1)V\,. \tag{11.43}$$

We find the boundary conditions for equation (11.43) by looking at (11.25) and evaluating at $y = 0$ and $y = 1$,

$$p_y|_{y=0} = \frac{1}{Re} V_{yy}(0)\,, \tag{11.44}$$

$$p_y|_{y=1} = \frac{1}{Re}(-\alpha^2 V_c + V_{yy}(1)) - (V_c)_t\,. \tag{11.45}$$

Equation (11.43) is nondynamic—it is an ODE in p with external forcing proportional to V. The solution to this ODE is

$$\begin{aligned} p = \frac{1}{\alpha}\Bigg\{ &\beta \int_0^y V(t,\eta)(2\eta-1)\sinh(\alpha(y-\eta))d\eta \\ &-\beta\frac{\cosh(\alpha y)}{\sinh(\alpha)}\int_0^1 V(t,\eta)(2\eta-1)\cosh(\alpha(1-\eta))d\eta \\ &-\frac{\cosh(\alpha(1-y))}{\sinh(\alpha)} p_y|_{y=0} + \frac{\cosh(\alpha y)}{\sinh(\alpha)} p_y|_{y=1}\Bigg\} \end{aligned} \tag{11.46}$$

$$\begin{aligned} = \frac{1}{\alpha}\Bigg\{ &\beta \int_0^y V(t,\eta)(2\eta-1)\sinh(\alpha(y-\eta))d\eta \\ &+\frac{\cosh(\alpha y)}{\sinh(\alpha)}\Bigg\{ -\beta\int_0^1 V(t,\eta)(2\eta-1)\cosh(\alpha(1-\eta))d\eta \\ &\qquad + \frac{1}{Re}(-\alpha^2 V_c + V_{yy}(1)) - (V_c)_t \Bigg\} \\ &-\frac{1}{Re}\frac{\cosh(\alpha(1-y))}{\sinh(\alpha)} V_{yy}(0)\Bigg\}\,. \end{aligned} \tag{11.47}$$

Before we substitute this expression into (11.38), we slightly modify it. This next step is similar to the shear beam design in Chapter 8. We design one controller to put the system into a strict feedback form so that we can use backstepping. Notice the V_c and $(V_c)_t$ terms in the equation. We use these to get rid of the unwanted integral term by setting $(V_c)_t$ to

$$(V_c)_t = \frac{1}{Re}\left(V_{yy}(1) - V_{yy}(0) - \alpha^2 V_c\right) - \beta \int_0^1 V(t,\eta)(2\eta - 1)\cosh(\alpha(1-\eta))d\eta \,. \tag{11.48}$$

This gives us the following expression:

$$p = \beta \int_0^y V(t,\eta)(2\eta - 1)\sinh(\alpha(y-\eta))d\eta + \frac{\cosh(\alpha y)}{\sinh(\alpha)}\frac{1}{Re}V_{yy}(0) - \frac{\cosh(\alpha(1-y))}{\sinh(\alpha)}\frac{1}{Re}V_{yy}(0) \,. \tag{11.49}$$

After substituting (11.49) into (11.38), we obtain

$$Y_t = \frac{1}{Re}(-\alpha^2 Y + Y_{yy}) + \frac{\beta}{2}y(y-1)Y - 8\pi j k_x(2y-1)V - \alpha\left\{\beta \int_0^y V(t,\eta)(2\eta - 1)\sinh(\alpha(y-\eta))d\eta + \frac{1}{Re}V_{yy}(0)\frac{\cosh(\alpha y) - \cosh(\alpha(1-y))}{\sinh(\alpha)}\right\} . \tag{11.50}$$

We can now use equations (11.33) and (11.37) to write the system in terms of Y and ω alone as follows:

$$Y_t = \frac{1}{Re}(-\alpha^2 Y + Y_{yy}) + \frac{\beta}{2}y(y-1)Y - 8\pi j k_x(2y-1)\int_0^y Y(\eta)d\eta - \alpha\left\{\beta \int_0^y \left(\int_0^\eta Y(\sigma)d\sigma\right)(2\eta - 1)\sinh(\alpha(y-\eta))d\eta + \frac{1}{Re}Y_y(0)\frac{\cosh(\alpha y) - \cosh(\alpha(1-y))}{\sinh(\alpha)}\right\} , \tag{11.51}$$

$$\omega_t = \frac{1}{Re}(-\alpha^2\omega + \omega_{yy}) + \frac{\beta}{2}y(y-1)\omega - 8\pi j k_z(2y-1)\int_0^y Y(\eta)d\eta \,. \tag{11.52}$$

By defining

$$\varepsilon = \frac{1}{Re} , \tag{11.53}$$

$$\phi(y) = \frac{\beta}{2}y(y-1) = 8\pi j k_x y(y-1) , \tag{11.54}$$

$$f(y,\eta) = 8j\left\{\pi k_x(2y-1) - 4\pi\frac{k_x}{\alpha}\sinh(\alpha(y-\eta)) - 2\pi k_x(2\eta-1)\cosh(\alpha(y-\eta))\right\} , \tag{11.55}$$

$$g(y) = \varepsilon\alpha^2\frac{\cosh(\alpha(1-y)) - \cosh(\alpha y)}{\sinh(\alpha)} , \tag{11.56}$$

$$h(y) = -8\pi k_z j(2y-1) , \tag{11.57}$$

we see that equations for Y and ω equations become

$$Y_t = \varepsilon(-\alpha^2 Y + Y_{yy}) + \phi(y)Y + g(y)Y_y(t,0) + \int_0^y f(y,\eta)Y(t,\eta)d\eta\,, \quad (11.58)$$

$$\omega_t = \varepsilon(-\alpha^2\omega + \omega_{yy}) + \phi(y)\omega + h(y)\int_0^y Y(\eta)d\eta\,, \quad (11.59)$$

with (11.40)–(11.42) still holding.

Equations (11.58), (11.59), and (11.40)–(11.42) form a plant model ready for the application of backstepping. Notice that we have a cascade system. The Y system is autonomous, while the ω system is nonautonomous and forced by the Y system. Also note that the Y system is unstable, while the ω system is stable without forcing (ϕ does not cause instability because it is a purely imaginary valued function; see Exercise 11.1). We use the following transformations to both decouple and stabilize the overall system:

$$\Psi = Y - \int_0^y K(y,\eta)Y(t,\eta)d\eta\,, \quad (11.60)$$

$$\Omega = \omega - \int_0^y \Gamma(y,\eta)Y(t,\eta)d\eta\,. \quad (11.61)$$

The first transformation eliminates the terms with g and f from the right-hand side of (11.58). The second transformation decouples Y and ω.

The target system is chosen as follows:

$$\Psi_t = \varepsilon(-\alpha^2\Psi + \Psi_{yy}) + \phi(y)\Psi\,, \quad (11.62)$$

$$\Omega_t = \varepsilon(-\alpha^2\Omega + \Omega_{yy}) + \phi(y)\Omega\,, \quad (11.63)$$

$$\Psi|_{y=0} = \Psi|_{y=1} = 0\,, \quad (11.64)$$

$$\Omega|_{y=1} = \Omega|_{y=1} = 0\,. \quad (11.65)$$

Both the Ψ and Ω systems are autonomous and stable. The terms $\phi(y)\Psi$ and $\phi(y)\Omega$ do not cause instability since the coefficient $\phi(y)$ is purely imaginary.

Since the PDEs for the gain kernels have been derived many times previously, we simply state the PDEs for K and Γ here without derivation (see Exercises 11.2 and 11.3):

$$\varepsilon K_{yy} = \varepsilon K_{\eta\eta} - f(y,\eta) + (\phi(\eta) - \phi(y))K + \int_\eta^y K(y,\xi)f(\xi,\eta)d\xi\,, \quad (11.66)$$

$$\varepsilon K(y,0) = \int_0^y K(y,\eta)g(\eta)d\eta - g(y)\,, \quad (11.67)$$

$$\varepsilon K(y,y) = -g(0)\,, \quad (11.68)$$

$$\varepsilon\Gamma_{yy} = \varepsilon\Gamma_{\eta\eta} - h(y) + (\phi(\eta) - \phi(y))\Gamma + \int_\eta^y \Gamma(y,\sigma)f(\sigma,\eta)d\sigma\,, \quad (11.69)$$

$$\varepsilon\Gamma(y,y) = 0\,, \quad (11.70)$$

$$\varepsilon\Gamma(y,0) = \int_0^y \Gamma(y,\eta)g(\eta)d\eta\,. \quad (11.71)$$

After solving for K and Γ (analytically or numerically), we obtain the controllers of the Y and ω systems from (11.60), (11.61) which are evaluated at $y = 1$:

$$Y(t,1) = \int_0^1 K(1,\eta) Y(t,\eta) d\eta \,, \tag{11.72}$$

$$\omega(t,1) = \int_0^1 \Gamma(1,\eta) Y(t,\eta) d\eta \,. \tag{11.73}$$

The ω controller needs only to use Y, as its purpose is only to decouple ω from Y. After substituting back into (11.35) and (11.36), the controllers for u and W become

$$U_c = \frac{-2\pi j}{\alpha^2}\Big(k_x Y(t,1) + k_z \omega(t,1)\Big) \,, \tag{11.74}$$

$$W_c = \frac{-2\pi j}{\alpha^2}\Big(k_z Y(y,1) - k_x \omega(t,1)\Big) \,, \tag{11.75}$$

which become

$$U_c = \int_0^1 \frac{4\pi^2}{\alpha^2}\Big(k_x K(1,\eta) + k_z \Gamma(1,\eta)\Big)\Big(k_x u(t,\eta) + k_z W(t,\eta)\Big) d\eta \,, \tag{11.76}$$

$$W_c = \int_0^1 \frac{4\pi^2}{\alpha^2}\Big(k_z K(1,\eta) - k_x \Gamma(1,\eta)\Big)\Big(k_x u(t,\eta) + k_z W(t,\eta)\Big) d\eta \tag{11.77}$$

by using (11.33) and (11.34).

Finally, we transform these controllers back into the physical space. To do so, we truncate the wavenumbers that we actuate at a certain number. This is feasible because high wavenumbers are already stable. The controllers in physical space are

$$\begin{aligned} V_c = & \int_0^t \int_{-\infty}^{\infty} \int_{-\infty}^{\infty} \int_{-\infty}^{\infty} \int_{-\infty}^{\infty} \frac{2\pi j}{Re} \chi(k_x, k_z) \\ & \times \Big\{ k_x \Big(u_y(\tau, \tilde{x}, \tilde{z}, 0) - u_y(\tau, \tilde{x}, \tilde{z}, 1) \Big) \\ & \qquad + k_z \Big(W_y(\tau, \tilde{x}, \tilde{z}, 0) - W_y(\tau, \tilde{x}, \tilde{z}, 1) \Big) \Big\} \\ & \times e^{\frac{\alpha^2}{Re}(t-\tau)} e^{2\pi j (k_x (x-\tilde{x}) + k_z (z - \tilde{z}))} dk_x dk_z d\tilde{x} d\tilde{z} d\tau \\ & - \int_0^t \int_0^1 \int_{-\infty}^{\infty} \int_{-\infty}^{\infty} \int_{-\infty}^{\infty} \int_{-\infty}^{\infty} V(\tau, \tilde{x}, \tilde{z}, \eta)(2\eta - 1) \\ & \times \chi(k_x, k_z) 16\pi k_x j \cosh(\alpha(1-\eta)) e^{\frac{\alpha^2}{Re}(t-\tau)} \\ & \times e^{2\pi j (k_x (x-\tilde{x}) + k_z (z-\tilde{z}))} dk_x dk_z d\tilde{x} d\tilde{z} d\eta d\tau \,, \end{aligned} \tag{11.78}$$

$$
\begin{aligned}
U_c = & \int_0^1 \int_{-\infty}^{\infty} \int_{-\infty}^{\infty} \int_{-\infty}^{\infty} \int_{-\infty}^{\infty} 4\pi^2 \frac{\chi(k_x, k_z)}{k_x^2 + k_z^2} \\
& \times \Big(k_x K(k_x, k_z, 1, \eta) + k_z \Gamma(k_x, k_z, 1, \eta) \Big) \\
& \times \Big(k_x u(t, \tilde{x}, \tilde{z}, \eta) + k_z W(t, \tilde{x}, \tilde{z}, \eta) \Big) \\
& \times e^{2\pi j (k_x (x - \tilde{x}) + k_z (z - \tilde{z}))} dk_x dk_z d\tilde{x} d\tilde{z} d\eta \,, \qquad (11.79) \\
W_c = & \int_0^1 \int_{-\infty}^{\infty} \int_{-\infty}^{\infty} \int_{-\infty}^{\infty} \int_{-\infty}^{\infty} 4\pi^2 \frac{\chi(k_x, k_z)}{k_x^2 + k_z^2} \\
& \times \Big(k_z K(k_x, k_z, 1, \eta) - k_x \Gamma(k_x, k_z, 1, \eta) \Big) \\
& \times \Big(k_x u(t, \tilde{x}, \tilde{z}, \eta) + k_z W(t, \tilde{x}, \tilde{z}, \eta) \Big) \\
& \times e^{2\pi j (k_x (x - \tilde{x}) + k_z (z - \tilde{z}))} dk_x dk_z d\tilde{x} d\tilde{z} d\eta \,, \qquad (11.80)
\end{aligned}
$$

where

$$
\chi(k_x, k_z) = \begin{cases} 1 & \begin{cases} \Big\{ |k_x| > m \text{ or } |k_z| > m \Big\} \\ \text{and} \\ \Big\{ |k_x| < M \text{ and } |k_x| < M \Big\}, \end{cases} \\ 0 & \text{else,} \end{cases} \qquad (11.81)
$$

and m and M are our cutoffs.

Physically, actuating small wavenumbers means that actuators can be spaced further apart, and the changes in space are relatively slow. Actuating high wavenumbers means that actuators spaced close together will change quickly in space.

As the ω subsystem in (11.59) is stable with $Y = 0$, one might ask why we use the transformation (11.73) and the controller (11.61) to decouple ω from Y. While this may seem unnecessary mathematically, there is an important practical reason for doing it. The coupling through $h(y)$ in (11.59) causes a large transient overshoot in the ω subsystem and is considered one of the principal causes of transition to turbulence, even at linearly stable Reynolds numbers [14]. The backstepping approach removes this harmful coupling between the Y (Orr–Sommerfeld) and ω (Squire) subsystems, yielding the first control law that directly targets the root cause of transition to turbulence.

11.4 Notes and References

Analysis and control of Navier–Stokes equations [1, 7, 13, 18, 20, 19, 24, 37, 39, 59, 60, 61, 64, 66, 72, 75, 78, 142, 146, 162, 163, 165, 170, 171, 172, 173, 180, 179] are exciting areas attracting the attention of those who seek some of the ultimate challenges in applied mathematics and control theory. A detailed overview of various efforts in controllability, optimal control, stabilization, and boundary control of Navier–Stokes is beyond the scope of this introductory book; however, we point out that good surveys reflecting the state of the art at the time of their publication are contained in [1, 24, 64].

A Lyapunov stabilization effort for Navier–Stokes equations was introduced in [1]; however, it was not until the application of backstepping to Navier–Stokes equations in [172, 173, 178, 37], which are references by which this chapter is inspired, that stabilization of the (nondiscretized) channel flow became possible for high Reynolds numbers.

Exercises

11.1. Using the Lyapunov function

$$\Lambda(t) = \frac{1}{2}\int_0^1 |\Omega|^2(t, y)dy\,,$$

where Ω is a complex-valued function and $|\cdot|$ denotes the modulus $(\Omega\bar{\Omega})$, show that the system

$$\begin{aligned}\Omega_t &= \varepsilon(-\alpha^2\Omega + \Omega_{yy}) + \phi(y)\Omega\,,\\ \Omega(0) &= 0\,,\\ \Omega(1) &= 0\end{aligned}$$

is exponentially stable when $\phi(y)$ is a purely imaginary valued function.

11.2. Consider the plant

$$Y_t = \varepsilon(-\alpha^2 Y + Y_{yy})\,, \tag{11.82}$$

$$\omega_t = \varepsilon(-\alpha^2\omega + \omega_{yy}) + h(y)\int_0^y Y(\eta)\,d\eta\,, \tag{11.83}$$

$$Y(0) = \omega(0) = 0\,. \tag{11.84}$$

Show that the transformation

$$\Omega(y) = \omega(y) - \int_0^y \Gamma(y, \eta)Y(\eta)d\eta$$

decouples Y and ω by transforming the plant (11.82)–(11.84) into the target system

$$\begin{aligned}Y_t &= \varepsilon(-\alpha^2 Y + Y_{yy})\,,\\ \Omega_t &= \varepsilon(-\alpha^2\Omega + \Omega_{yy})\,,\\ Y(0) &= \Omega(0) = 0\end{aligned}$$

when the following PDE for Γ is satisfied:

$$\begin{aligned}\varepsilon\Gamma_{yy}(y, \eta) &= \varepsilon\Gamma_{\eta\eta}(y, \eta) - h(y)\,,\\ \Gamma(y, 0) &= 0\,,\\ \Gamma(y, y) &= 0\,.\end{aligned}$$

11.3. Show that the transformation

$$\Psi(y) = Y(y) - \int_0^y K(y,\eta)Y(\eta)\,d\eta$$

transforms the plant

$$\begin{aligned} Y_t &= \varepsilon(-\alpha^2 Y + Y_{yy}) + \phi(y)Y + g(y)Y_y(0)\,, \\ Y(0) &= 0 \end{aligned}$$

into the target system

$$\begin{aligned} \Psi_t &= \varepsilon(-\alpha^2 \Psi + \Psi_{yy}) + \phi(y)\Psi\,, \\ \Psi(0) &= 0 \end{aligned}$$

when the following PDE for K is satisfied:

$$\begin{aligned} \varepsilon K_{yy} &= \varepsilon K_{\eta\eta} + (\phi(\eta) - \phi(y))K\,, \\ \varepsilon K(y,0) &= \int_0^y K(y,\eta)g(\eta)d\eta - g(y)\,, \\ \varepsilon K(y,y) &= -g(0)\,. \end{aligned}$$

11.4. The double backstepping transformation

$$\Psi(y) = Y(y) - \int_0^y K(y,\eta)Y(\eta)\,d\eta, \tag{11.85}$$

$$\Omega(y) = \omega(y) - \int_0^y \Gamma(y,\eta)Y(\eta)\,d\eta, \tag{11.86}$$

with inverse

$$Y(y) = \Psi(y) + \int_0^y L(y,\eta)\Psi(\eta)\,d\eta\,, \tag{11.87}$$

$$\omega(y) = \Omega(y) + \int_0^y \Theta(y,\eta)\Psi(\eta)\,d\eta\,, \tag{11.88}$$

transforms the plant

$$\begin{aligned} Y_t &= \varepsilon(-\alpha^2 Y + Y_{yy}) + \phi(y)Y + g(y)Y_y(0) + \int_0^y f(y,\eta)Y(\eta)d\eta\,, \\ \omega_t &= \varepsilon(-\alpha^2 \omega + \omega_{yy}) + \phi(y)\omega + h(y)\int_0^y Y(\eta)d\eta\,, \\ Y(0) &= \omega(0) = 0 \end{aligned}$$

into the target system

$$\begin{aligned} \Psi_t &= \varepsilon(-\alpha^2 \Psi + \Psi_{yy}) + \phi(y)\Psi, \\ \Omega_t &= \varepsilon(-\alpha^2 \Omega + \Omega_{yy}) + \phi(y)\Omega\,, \\ \Psi(0) &= \Omega(0) = 0\,. \end{aligned}$$

Show that the kernel of the inverse transformation (11.88) can be computed as

$$\Theta(y, \eta) = \Gamma(y, \eta) + \int_{\eta}^{y} \Gamma(y, \sigma) L(\sigma, \eta) d\sigma$$

by first plugging (11.87) into (11.86), then plugging the modified (11.61) into (11.88), and finally using formula (4.36).

Chapter 12

Motion Planning for PDEs

This book is almost entirely devoted to stabilization problems for PDEs. As such, the book deals with the design of *feedback* laws. However, since the PDEs we address in this book are linear, the same feedback laws designed for stabilization of zero equilibria are capable of also stabilizing any other feasible trajectories of these PDEs. To accomplish such tasks, we need to be able to first generate such trajectories as functions of t and x. In other words, we need to be able to generate not only the necessary feedback controls but also the *feedforward* controls.

This chapter provides an introduction to some recently emerging ideas in *trajectory generation*, or *motion planning*, or, simply put, open-loop control for PDEs.

As we shall see, once we are able to generate a *reference* trajectory for the state of the PDE, we can easily combine our feedforward and feedback controls to stabilize this trajectory. For example, if the control input to the PDE is of Dirichlet type $u(1,t)$ and we have designed a feedback control

$$u(1,t) = \int_0^1 k(1,y)u(y,t)\,dy\,,$$

and if we have generated a state reference trajectory $u^r(x,t)$, then the overall "feedback+feedforward" control law

$$u(1,t) = u^r(1,t) + \int_0^1 k(1,y)(u(y,t) - u^r(y,t))\,dy$$

will (exponentially) stabilize the trajectory, namely, it will ensure that $u(x,t) - u^r(x,t) \to 0$ as $t \to \infty$ (for all x, in an appropriate sense).

While trajectory tracking for a *state* reference is a relevant problem, a truly important engineering problem is that of tracking an *output* reference. To solve such a problem, one starts from an output reference trajectory, for example, $u^r(0,t)$, as the desired temporal waveform of the system output $u(0,t)$ at $x = 0$; generates the state trajectory $u^r(x,t)$ for all x (including $x = 1$, which produces the *control reference*); and then combines this result with a feedback control law to stabilize the trajectory $u^r(x,t)$ and to force the output $u(0,t)$ to track the output trajectory $u^r(0,t)$. This chapter focuses on the generation of the state trajectory $u^r(x,t)$ given the output trajectory $u^r(0,t)$, which is often referred to as motion planning.

Our coverage in this chapter is driven not by a desire to achieve generality but to achieve clarity. Instead of trying to find a state trajectory for an arbitrary output trajectory, we go through a series of examples with reference outputs common in practice—exponential, polynomial, and sinusoidal signals—for a series of different PDEs which includes the heat equation, the reaction-diffusion equation, the wave equation, the Euler–Bernoulli beam, and a first-order hyperbolic PDE. We also consider different types of outputs, which include the "Dirichlet" outputs $u(0,t)$ and the "Neumann" outputs $u_x(0,t)$.

12.1 Trajectory Generation

Example 12.1 As our first example, we consider the plant

$$u_t = u_{xx}\,, \tag{12.1}$$

$$u_x(0) = 0 \tag{12.2}$$

with reference output

$$u^r(0,t) = 1 + 2t - t^2\,. \tag{12.3}$$

In a physical sense, we want to generate a temperature trajectory at $x = 1$ such that the temperature evolution at $x = 0$ is given by (12.3). To find the reference input $u^r(1,t)$ we first need to construct the full-state trajectory $u^r(x,t)$, which simultaneously satisfies (12.1), (12.2), and (12.3). Let us search for the state trajectory in the following form:

$$u^r(x,t) = \sum_{k=0}^{\infty} a_k(t)\frac{x^k}{k!}\,. \tag{12.4}$$

This is a Taylor series in x with time-varying coefficients $a_k(t)$ that need to be determined from (12.1)–(12.3). From (12.3)–(12.4) we see that

$$a_0(t) = u^r(0,t) = 1 + 2t - t^2. \tag{12.5}$$

The boundary condition (12.2) gives

$$a_1(t) = u_x^r(0,t) = 0\,. \tag{12.6}$$

The next step is to substitute (12.4) into (12.1) as follows:

$$\begin{aligned}\sum_{k=0}^{\infty} \dot{a}_k(t)\frac{x^k}{k!} &= \frac{\partial^2}{\partial x^2}\sum_{k=0}^{\infty} a_k(t)\frac{x^k}{k!}\\ &= \sum_{k=2}^{\infty} a_k(t)\frac{k(k-1)x^{k-2}}{k!}\\ &= \sum_{k=2}^{\infty} a_k(t)\frac{x^{k-2}}{(k-2)!}\\ &= \sum_{k=0}^{\infty} a_{k+2}(t)\frac{x^k}{k!}.\end{aligned} \tag{12.7}$$

We get the recursive relationship

$$a_{k+2}(t) = \dot{a}_k(t)\,. \tag{12.8}$$

Using (12.6) and (12.5) with (12.8) results in

$$\begin{aligned} a_0 &= 1 + 2t - t^2, & a_1 &= 0\,, \\ a_2 &= 2 - 2t, & a_3 &= 0\,, \\ a_4 &= -2, & a_5 &= 0\,, \\ a_6 &= 0, & a_i &= 0 \quad \text{for } i > 6. \end{aligned}$$

This gives the reference state trajectory

$$u^r(x,t) = 1 + 2t + t^2 + (1-t)x^2 - \frac{1}{12}x^4$$

and the input signal

$$u^r(1,t) = \frac{23}{12} + t - t^2. \quad \diamond$$

Note that the output matches the reference output trajectory only if the initial condition of the plant is satisfied by the state trajectory, that is, if $u(x,0) = 1 + x^2 - \frac{1}{12}x^4$. To asymptotically track the reference signal as $t \to \infty$ for an arbitrary initial condition, we need to apply feedback.

The basic idea introduced in Example 12.1 is to use the "spatial Taylor series" representation (12.4) and to find the temporal coefficients $a_k(t)$ using a recursion such as (12.8). This idea will permeate our developments in the subsequent examples. However, while in Example 12.1 the Taylor series was finite, in the subsequent examples the series will be infinite and will have to be summed.

Example 12.2 Consider the reaction-diffusion equation

$$u_t = u_{xx} + \lambda u\,, \tag{12.9}$$

$$u_x(0) = 0 \tag{12.10}$$

with a reference output

$$u^r(0,t) = \mathrm{e}^{\alpha t}\,. \tag{12.11}$$

Let us find the reference input $u^r(1,t)$. We again postulate the full-state reference trajectory in the form

$$u^r(x,t) = \sum_{k=0}^{\infty} a_k(t)\frac{x^k}{k!}\,. \tag{12.12}$$

From (12.11) and the boundary condition (12.10), we have

$$a_0(t) = \mathrm{e}^{\alpha t}, \qquad a_1(t) = 0,$$

and from the PDE (12.9), we get

$$a_{k+2}(t) = \dot{a}_k(t) - \lambda a_k(t).$$

These conditions give

$$\begin{aligned} a_{2k+1} &= 0\,, \\ a_{2k+2} &= \dot{a}_{2k} - \lambda a_{2k}\,, \\ a_2 &= (\alpha - \lambda)e^{\alpha t}\,, \\ a_4 &= (\alpha - \lambda)^2 e^{\alpha t}\,, \\ a_{2k} &= (\alpha - \lambda)^k e^{\alpha t} \end{aligned}$$

for $k = 0, 1, 2, \ldots,$ so that the state trajectory is

$$\begin{aligned} u^r(x,t) &= \sum_{k=0}^{\infty} (\alpha - \lambda)^k e^{\alpha t} \frac{x^{2k}}{(2k)!} \\ &= \sum_{k=0}^{\infty} e^{\alpha t} \frac{(\sqrt{\alpha - \lambda}\,x)^{2k}}{(2k)!} \\ &= e^{\alpha t} \begin{cases} \cosh(\sqrt{\alpha - \lambda}\,x) & \alpha \geq \lambda\,, \\ \cos(\sqrt{\lambda - \alpha}\,x) & \alpha < \lambda\,. \end{cases} \end{aligned} \tag{12.13}$$

The reference input is

$$u^r(1,t) = e^{\alpha t} \begin{cases} \cosh(\sqrt{\alpha - \lambda}) & \alpha \geq \lambda\,, \\ \cos(\sqrt{\lambda - \alpha}) & \alpha < \lambda\,. \end{cases} \qquad \diamond$$

Example 12.2 was the first to introduce the infinite recursion and summation (12.13). This example dealt with an exponential reference function. The next few examples will deal with sinusoidal reference functions.

The following formulae are useful when calculating the trajectories for sinusoidal reference outputs in the upcoming examples:

$$\begin{aligned} \cosh(a) &= \sum_{k=0}^{\infty} \frac{a^{2k}}{(2k)!}\,, & \sinh(a) &= \sum_{k=0}^{\infty} \frac{a^{2k+1}}{(2k+1)!}\,, \\ \cosh(ja) &= \cos(a)\,, & \sinh(ja) &= j\sin(a)\,, \\ \cos(ja) &= \cosh(a)\,, & \sin(ja) &= j\sinh(a)\,. \end{aligned}$$

Example 12.3 Consider the plant

$$u_t = u_{xx}\,, \tag{12.14}$$

$$u_x(0) = 0 \tag{12.15}$$

with reference output

$$u^r(0,t) = \sin(\omega t)\,. \tag{12.16}$$

Since $\sin(\omega t) = \text{Im}\{e^{j\omega t}\}$, we can get the reference trajectory by setting $\lambda = 0$ and $\alpha = j\omega$ in the previous example. We have

$$u^r(x,t) = \text{Im}\left\{\cosh(\sqrt{j\omega}x)e^{j\omega t}\right\}. \tag{12.17}$$

Using the identity $\sqrt{j} = (1+j)/2$, we get

$$\begin{aligned} u^r(x,t) &= \text{Im}\left\{\cosh\left((1+j)\sqrt{\frac{\omega}{2}x}\right)e^{j\omega t}\right\} \\ &= \text{Im}\left\{\frac{e^{\sqrt{\frac{\omega}{2}}x+j\left(\omega t+\sqrt{\frac{\omega}{2}}x\right)}+e^{-\sqrt{\frac{\omega}{2}}x+j\left(\omega t-\sqrt{\frac{\omega}{2}}x\right)}}{2}\right\} \\ &= \frac{1}{2}e^{\sqrt{\frac{\omega}{2}}x}\sin\left(\omega t+\sqrt{\frac{\omega}{2}}x\right)+\frac{1}{2}e^{-\sqrt{\frac{\omega}{2}}x}\sin\left(\omega t-\sqrt{\frac{\omega}{2}}x\right). \end{aligned} \tag{12.18}$$

Finally, the reference input is

$$u^r(1,t) = \frac{1}{2}e^{\sqrt{\frac{\omega}{2}}}\sin\left(\omega t+\sqrt{\frac{\omega}{2}}\right)+\frac{1}{2}e^{-\sqrt{\frac{\omega}{2}}}\sin\left(\omega t-\sqrt{\frac{\omega}{2}}\right). \tag{12.19}$$

◇

All the examples so far have dealt with Dirichlet-type outputs $u(0,t)$. The next example deals with a Neumann-type output, $u_x(0,t)$.

Example 12.4 Consider the plant

$$u_t = u_{xx}, \tag{12.20}$$

$$u(0) = 0 \tag{12.21}$$

with the reference output

$$u_x^r(1) = \sin(\omega t). \tag{12.22}$$

Postulating $u^r(x,t)$ in the form of (12.12), we get $a_{i+2} = \dot{a}_i$, and the boundary condition gives

$$a_{2k} = 0, \tag{12.23}$$

$$a_{2k+1} = a_1^{(k)}. \tag{12.24}$$

The state trajectory becomes

$$u^r(x,t) = \sum_{k=0}^{\infty} a_1^{(k)}(t)\frac{x^{2k+1}}{(2k+1)!}. \tag{12.25}$$

The output reference is

$$u_x^r(1,t) = \sum_{k=0}^{\infty}\frac{a_1^{(k)}(t)}{(2k)!} = \sin(\omega t) = \text{Im}\{e^{j\omega t}\}. \tag{12.26}$$

Suppose that

$$a_1(t) = \mathrm{Im}\{Ae^{j\omega t}\},$$

where A is a constant to be determined. Then

$$a_1^{(k)}(t) = \mathrm{Im}\{A(j\omega)^k e^{j\omega t}\}.$$

From (12.26) we get

$$\mathrm{Im}\left\{Ae^{j\omega t}\sum_{k=0}^{\infty}\frac{(\sqrt{j\omega})^{2k}}{(2k)!}\right\} = \mathrm{Im}\{e^{j\omega t}\}, \tag{12.27}$$

$$Ae^{j\omega t}\cosh(\sqrt{j\omega}) = e^{j\omega t} \tag{12.28}$$

so that $A = 1/\cosh(\sqrt{j\omega})$. The state trajectory is now

$$u^r(x,t) = \sum_{k=0}^{\infty} a_1^{(k)}(t)\frac{x^{2k+1}}{(2k+1)!} \tag{12.29}$$

$$= \mathrm{Im}\left\{A\sum_{k=0}^{\infty}(j\omega)^k\frac{x^{2k+1}}{(2k+1)!}e^{j\omega t}\right\} \tag{12.30}$$

$$= \mathrm{Im}\left\{\frac{A}{\sqrt{j\omega}}\sum_{k=0}^{\infty}\frac{(\sqrt{j\omega}x)^{2k+1}}{(2k+1)!}e^{j\omega t}\right\} \tag{12.31}$$

$$= \mathrm{Im}\left\{\frac{\sinh(\sqrt{j\omega}x)}{\sqrt{j\omega}\cosh(\sqrt{j\omega})}e^{j\omega t}\right\}, \tag{12.32}$$

which gives the reference input

$$u^r(1,t) = \mathrm{Im}\left\{\frac{\tanh(\sqrt{j\omega})}{\sqrt{j\omega}}e^{j\omega t}\right\}. \tag{12.33}$$

◇

Remark 12.5. If in the above example the output reference is exponential,

$$u_x^r(1,t) = e^{\alpha t}, \quad \alpha \in \mathbb{R}, \tag{12.34}$$

then the corresponding input reference is

$$u^r(1,t) = \frac{e^{\alpha t}}{\sqrt{|\alpha|}}\begin{cases}\tanh(\sqrt{|\alpha|}), & \alpha > 0,\\ \tan(\sqrt{|\alpha|}), & \alpha < 0.\end{cases} \tag{12.35}$$

All the examples so far dealt with parabolic PDEs. The remaining examples deal with hyperbolic PDEs.

Example 12.6 Consider the wave equation

$$u_{tt} = u_{xx}, \tag{12.36}$$

$$u(0) = 0 \tag{12.37}$$

with the reference output

$$u_x^r(0,t) = \sin(\omega t)\,. \tag{12.38}$$

Searching for $u^r(x,t)$ in the form of (12.12), we get

$$a_0 = 0,\quad a_1(t) = \sin(\omega t) = \mathrm{Im}\{e^{j\omega t}\}\,,$$
$$a_{i+2} = \ddot{a}_i(t)\,,$$

which gives

$$a_{2k} = 0\,,$$
$$a_{2k+1}(t) = (j\omega)^{2k}\mathrm{Im}\{e^{j\omega t}\}\,.$$

The state reference becomes

$$\begin{aligned} u^r(x,t) &= \mathrm{Im}\left\{\frac{e^{j\omega t}}{j\omega}\sum_{k=0}^{\infty}\frac{(j\omega x)^{2k+1}}{(2k+1)!}\right\} \\ &= \mathrm{Im}\left\{\frac{e^{j\omega t}}{j\omega}\sinh(j\omega x)\right\} \\ &= \mathrm{Im}\left\{\frac{e^{j\omega t}}{\omega}\sin(\omega x)\right\} \\ &= \frac{1}{\omega}\sin(\omega x)\sin(\omega t), \end{aligned}$$

and the reference input is

$$u^r(1,t) = \frac{\sin(\omega)}{\omega}\sin(\omega t)\,. \tag{12.39}$$

Note that, for the same desired reference output trajectory, the reference input for the heat equation (Example 12.3) has a much more complicated form. ◇

Example 12.7 Consider a wave equation with Kelvin–Voigt damping,

$$\varepsilon u_{tt} = (1 + d\partial_t)u_{xx}\,, \tag{12.40}$$
$$u_x(0) = 0\,, \tag{12.41}$$

and with the reference output

$$u^r(0,t) = \sin(\omega t)\,. \tag{12.42}$$

Following the procedure in the previous examples, one can obtain the following reference state trajectory:

$$\begin{aligned} u^r(x,t) = \frac{1}{2}\Bigg[& \mathrm{e}^{\sqrt{\varepsilon}\frac{\omega\sqrt{\sqrt{1+\omega^2d^2}-1}}{\sqrt{2}\sqrt{1+\omega^2d^2}}x}\sin\left(\omega\left(t+\sqrt{\varepsilon}\frac{\omega\sqrt{\sqrt{1+\omega^2d^2}+1}}{\sqrt{2}\sqrt{1+\omega^2d^2}}x\right)\right) \\ & + \mathrm{e}^{-\sqrt{\varepsilon}\frac{\omega\sqrt{\sqrt{1+\omega^2d^2}-1}}{\sqrt{2}\sqrt{1+\omega^2d^2}}x}\sin\left(\omega\left(t-\sqrt{\varepsilon}\frac{\omega\sqrt{\sqrt{1+\omega^2d^2}+1}}{\sqrt{2}\sqrt{1+\omega^2d^2}}x\right)\right)\Bigg]\,. \end{aligned} \tag{12.43}$$

◇

Example 12.8 Consider the Euler–Bernoulli beam model,

$$u_{tt} + u_{xxxx} = 0\,, \tag{12.44}$$

$$u_{xx}(0) = u_{xxx}(0) = 0\,, \tag{12.45}$$

with the reference outputs

$$u^r(0,t) = \sin(\omega t)\,, \tag{12.46}$$

$$u^r_x(0,t) = 0\,. \tag{12.47}$$

Note that, because the beam equation is fourth order in x, we are free to assign two independent reference outputs and to choose two reference inputs to generate the two independent outputs.

Searching for $u^r(x,t)$ in the form of (12.12), we get

$$a_0 = \sin(\omega t) = \operatorname{Im}\left\{e^{j\omega t}\right\}\,, \tag{12.48}$$

$$a_1 = a_2 = a_3 = 0\,, \tag{12.49}$$

$$a_{i+4} = -\ddot{a}_i\,. \tag{12.50}$$

Therefore,

$$a_{4k} = (-1)^k a_0^{(2k)} = \operatorname{Im}\left\{(-1)^k (j\omega)^{2k} e^{j\omega t}\right\} = \omega^{2k}\sin(\omega t)\,, \tag{12.51}$$

$$a_{4k+1} = a_{4k+2} = a_{4k+3} = 0\,. \tag{12.52}$$

The reference trajectory becomes

$$\begin{aligned} u^r(x,t) &= \sum_{k=0}^{\infty} \omega^{2k} \frac{x^{4k}}{(4k)!} \sin(\omega t) \\ &= \sum_{k=0}^{\infty} \frac{(\sqrt{\omega}x)^{4k}}{(4k)!} \sin(\omega t) \\ &= \frac{1}{2}[\cosh(\sqrt{\omega}x) + \cos(\sqrt{\omega}x)]\sin(\omega t)\,, \end{aligned}$$

and reference inputs are

$$\begin{aligned} u^r(1,t) &= \frac{1}{2}[\cosh(\sqrt{\omega}) + \cos(\sqrt{\omega})]\sin(\omega t)\,, \\ u^r_x(1,t) &= \frac{\sqrt{\omega}}{2}[\sinh(\sqrt{\omega}x) - \sin(\sqrt{\omega}x)]\sin(\omega t)\,. \end{aligned} \quad \diamond$$

Example 12.9 Consider the first-order hyperbolic PDE

$$u_t = u_x + g u(0) \tag{12.53}$$

with the reference output

$$u^r(0,t) = \sin(\omega t)\,. \tag{12.54}$$

Searching for the reference trajectory in the form of (12.12), we get

$$\begin{aligned} a_0 &= \sin(\omega t) = \operatorname{Im}\left\{e^{j\omega t}\right\}, \\ a_1 &= \dot a_0 - g u(0) = \operatorname{Im}\left\{(j\omega - g)e^{j\omega t}\right\}, \\ a_{k+1} &= \dot a_k = \operatorname{Im}\left\{(j\omega - g)(j\omega)^k e^{j\omega t}\right\} \\ &= \operatorname{Im}\left\{\left(1 - \frac{g}{j\omega}\right)(j\omega)^{k+1} e^{j\omega t}\right\}. \end{aligned}$$

The reference trajectory becomes

$$\begin{aligned} u^r(x,t) &= \operatorname{Im}\left\{e^{j\omega t} + \left(1 - \frac{g}{j\omega}\right)\sum_{k=1}^{\infty}\frac{(j\omega x)^k}{k!}e^{j\omega t}\right\} \\ &= \operatorname{Im}\left\{\left[\frac{g}{j\omega} + \left(1 - \frac{g}{j\omega}\right)e^{j\omega x}\right]e^{j\omega t}\right\} \\ &= -\frac{g}{\omega}[\cos(\omega t) - \cos(\omega(t+x))] + \sin(\omega(t+x)), \end{aligned}$$

which gives the reference input

$$u^r(1,t) = \frac{g}{\omega}[\cos(\omega(t+1)) - \cos(\omega t)] + \sin(\omega(t+1)). \tag{12.55}$$

◇

12.2 Trajectory Tracking

The reference inputs obtained in the previous section achieve the desired output from only one particular initial condition. One can asymptotically track the reference signal for an arbitrary initial condition by using the feedback controllers developed in the previous chapters to stabilize the reference trajectory.

Example 12.10 Consider the plant

$$u_t = u_x + g u(0).$$

Suppose we want the output $u(0,t)$ to track the reference trajectory $u^r(0,t) = \sin(\omega t)$ using $u(1,t)$ as the input. In Example 12.9 we obtained the following state reference trajectory:

$$u^r(x,t) = \frac{g}{\omega}[\cos(\omega(t+x)) - \cos(\omega t)] + \sin(\omega(t+x)). \tag{12.56}$$

Let us introduce the error variable

$$\tilde u(x,t) = u(x,t) - u^r(x,t); \tag{12.57}$$

then we have the equation

$$\tilde u_t = \tilde u_x + g\tilde u(0),$$

which we want to regulate to zero. In Chapter 9 we derived the stabilizing controller

$$\tilde{u}(1) = -\int_0^1 g e^{g(1-y)} \tilde{u}(y)\, dy\,. \tag{12.58}$$

Using the definition of $\tilde{u}(x, t)$, we get

$$u(1, t) = u^r(1, t) - \int_0^1 g e^{g(1-y)} [u(y, t) - u^r(y, t)]\, dy \tag{12.59}$$

$$= u^r(1, t) + \int_0^1 g e^{g(1-y)} u^r(y, t)\, dy - \int_0^1 g e^{g(1-y)} u(y, t)\, dy \tag{12.60}$$

$$= \frac{g}{\omega}[\cos(\omega(t+1)) - \cos(\omega t)] + \sin(\omega(t+1)) \tag{12.61}$$

$$+ \int_0^1 g e^{g(1-y)} \Big[\frac{g}{\omega}\cos(\omega(t+y)) - \cos(\omega t) + \sin(\omega(t+x))\Big] dy \tag{12.62}$$

$$- \int_0^1 g e^{g(1-y)} u(y, t)\, dy \tag{12.63}$$

$$= \frac{g}{\omega}[\cos(\omega(t+1)) - \cos(\omega t)] + \sin(\omega(t+1))$$

$$- \frac{g}{\omega}[\cos(\omega(t+1)) - \cos(\omega t)] \tag{12.64}$$

$$- \int_0^1 g e^{g(1-y)} u(y, t) dy\,. \tag{12.65}$$

Hence, we obtain the complete control law (feedforward plus feedback) as

$$u(1, t) = \sin(\omega(t+1)) - \int_0^1 g e^{g(1-y)} u(y, t) dy\,. \tag{12.66}$$

◇

Remark 12.11. Note that the controller above has the form of a pure "unit" advance of the reference trajectory $u^r(0, t) = \sin(\omega t)$ plus feedback, where the pure advance comes from the target system used in the control design. This is not an accident. It turns out that the backstepping controller always has the structure of "reference input designed for the target system plus feedback." This greatly simplifies the design for complicated plants since the target system is usually much simpler than the original plant (see Exercise 12.5).

Example 12.12 Figure 12.1 displays the trajectory tracking results for the open-loop trajectory designed in Example 12.7, and for a feedback law from Section 7.3, for the *almost periodic* trajectory given by

$$u^r(0, t) = \frac{1}{2}\left[\sin\left(\frac{\pi}{2}t\right) + \sin\left(\frac{\sqrt{2}\pi}{2}t\right)\right]. \tag{12.67}$$

◇

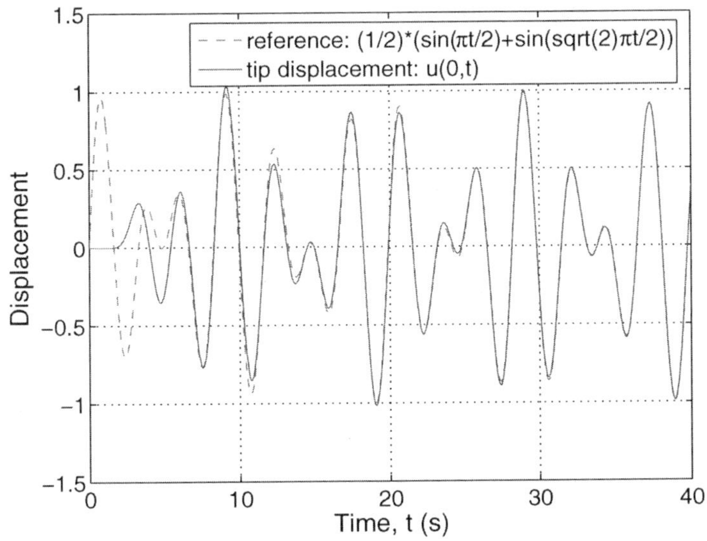

Figure 12.1. *Asymptotic tracking for the wave equation with Kelvin–Voigt damping.*

12.3 Notes and References

The foundations for motion planning for several classes of PDEs were laid in the late 1990s in papers by Rouchon and coworkers [50, 56, 105, 147], Laroche and Martin [104], Fliess and Mounier [55], Meurer and Zeitz [128], Ollivier and Sedoglavic [134], and Petit and Del Vecchio [138]. These results extended the concept of "flatness" from finite-dimensional to infinite-dimensional systems through the parameterizations of trajectories via Gevrey functions and advance/delay operations on the reference trajectory. In Section 12.1 we specialized some of these ideas to particular classes of trajectories.

An important new idea for motion planning and trajectory stabilization for systems more complex than simple heat, wave, or beam equations is introduced in Remark 12.11 and Exercise 12.5.

Exercises

12.1. For the heat equation

$$u_t = u_{xx} ,$$
$$u_x(0) = 0 ,$$

find the input reference signal $u^r(1, t)$ so that the output $u(0, t)$ obeys the reference signal

$$u^r(0, t) = t^3 .$$

12.2. For the heat equation

$$u_t = u_{xx}\,,$$
$$u(0) = 0\,,$$

find the input reference signal $u^r(1,t)$ so that the output $u_x(0,t)$ obeys the reference signal

$$u_x^r(0,t) = \sin \omega t\,.$$

12.3. For the Euler–Bernoulli beam

$$u_{tt} + u_{xxxx} = 0\,,$$
$$u_{xx}(0) = 0\,,$$
$$u_{xxx}(0) = 0\,,$$

show that

$$u^r(x,t) = \frac{\sinh\left(\sqrt{\omega}x\right) + \sin\left(\sqrt{\omega}x\right)}{2\sqrt{\omega}} \sin(\omega t)$$

is a solution to the system. This result shows that you can produce the output trajectory

$$u^r(0,t) = 0\,,$$
$$u_x^r(0,t) = \sin(\omega t)$$

with the controls

$$u^r(1,t) = \frac{\sinh\sqrt{\omega} + \sin\sqrt{\omega}}{2\sqrt{\omega}} \sin(\omega t)\,,$$
$$u_x^r(1,t) = \frac{\cosh\sqrt{\omega} + \cos\sqrt{\omega}}{2} \sin(\omega t)\,.$$

12.4. Consider the undamped wave equation

$$u_{tt} = u_{xx}\,,$$
$$u_x(0) = 0\,.$$

Find the state reference trajectory $u^r(x,t)$ that corresponds to the output reference

$$u^r(0,t) = \sin \omega t\,.$$

Then, recalling that

$$u_x(1,t) = -c_0 u(1,t) - c_1\left(u_t(1,t) + c_0\int_0^1 u_t(y,t)dy\right), \qquad c_0, c_1 > 0\,,$$

is a stabilizing controller, find the functions $M(\omega, c_0, c_1)$ and $\phi(\omega, c_0, c_1)$ to ensure that the controller

$$u_x(1,t) = M(\omega, c_0, c_1)\sin\left(\omega t + \phi(\omega, c_0, c_1)\right)$$
$$-c_0 u(1,t) - c_1\left(u_t(1,t) + c_0\int_0^1 u_t(y,t)dy\right)$$

guarantees that the output $u(0,t)$ achieves asymptotic tracking of the output reference $u^r(0,t) = \sin \omega t$.

Hint: First, show that

$$M(\omega, c_0, c_1)\sin(\omega t + \phi(\omega, c_0, c_1)) = u_x^r(1,t) + c_0 u^r(1,t) + c_1\left(u_t^r(1,t) + c_0\int_0^1 u_t^r(y,t)dy\right).$$

Note that the left side of this expression is much more suitable for online implementation because M and ϕ can be precomputed, whereas the implementation on the right requires the integration of the reference signal to be done online.

12.5. This exercise presents an alternative approach to doing trajectory tracking, when compared to the approach in Example 12.2. Consider the reaction-diffusion equation

$$u_t = u_{xx} + \lambda(x)u\,,$$
$$u_x(0) = 0\,.$$

When $\lambda(x)$ is spatially varying, the motion planning procedure cannot produce a closed-form solution, even for basic output trajectories such as $u^r(0,t) = \sin \omega t$ or $u^r(0,t) = \mathrm{e}^{\alpha t}$. (In fact, even for the case $\lambda = \text{const} \neq 0$, the reference trajectory becomes considerably more complicated than the trajectory for $\lambda = 0$.) However, if our objective is just tracking, namely, not trajectory generation per se but finding a feedback law that stabilizes a trajectory that corresponds to a certain o utput reference, then it turns out that the feedback law

$$u(1,t) = \frac{1}{2}\left[\mathrm{e}^{\sqrt{\frac{\omega}{2}}}\sin\left(\omega t + \sqrt{\frac{\omega}{2}}\right) + \mathrm{e}^{-\sqrt{\frac{\omega}{2}}}\sin\left(\omega t - \sqrt{\frac{\omega}{2}}\right)\right] + \int_0^1 k(1,y)u(y,t)dy\,,$$

where $k(x,y)$ is the solution of the kernel PDE

$$k_{xx}(x,y) - k_{yy}(x,y) = \lambda(y)k(x,y)\,,$$
$$k_y(x,0) = 0\,,$$
$$k(x,x) = -\frac{1}{2}\int_0^x \lambda(y)dy\,,$$

guarantees that

$$u(x,t) - \int_0^x k(x,y)u(y,t)dy - \frac{1}{2}\left[\mathrm{e}^{\sqrt{\frac{\omega}{2}}x}\sin\left(\omega t + \sqrt{\frac{\omega}{2}}x\right) + \mathrm{e}^{-\sqrt{\frac{\omega}{2}}x}\sin\left(\omega t - \sqrt{\frac{\omega}{2}}x\right)\right] \to 0$$

as $t \to \infty$ for all $x \in [0,1]$, which means, in particular, that asymptotic tracking of the reference output $u^r(0,t) = \sin \omega t$ is achieved, namely,

$$u(0,t) - \sin \omega t \to 0 \quad \text{as } t \to \infty\,.$$

Explain (prove) this result. It is helpful to use the following notation:

$$w(x,t) = u(x,t) - \int_0^x k(x,y)u(y,t)dy\,,$$
$$w^r(x,t) = u^r(x,t) - \int_0^x k(x,y)u^r(y,t)dy\,,$$
$$\tilde{w}(x,t) = w(x,t) - w^r(x,t)$$

and note that all the three w-variables, w, w^r, and $\tilde{w}$, satisfy a heat equation with a Neumann boundary condition at $x = 0$ and with a boundary condition given by

$$\tilde{w}(1,t) = w(1,t) - w^r(1,t) = 0\,, \tag{12.68}$$

which determines the control law.

The point of this exercise is that trajectory tracking can be pursued for complicated, spatially varying, parabolic, and hyperbolic PDEs if one uses backstepping for trajectory stabilization. It is sufficient to develop trajectory generation for the heat or wave equation and add the input (labeled in this exercise as $w^r(1,t)$) to the usual stabilizing backstepping feedback law.

Chapter 13

Adaptive Control for PDEs

In applications that incorporate thermal, fluid, or chemically reacting dynamics, physical parameters are often unknown. Thus a need exists for a "parameter-adaptive" technique which estimates such unknown parameters, continuously recomputes the controller gains, and applies the resulting controller to stabilize a potentially unstable, parametrically uncertain plant. Such an objective is incompatible with methods that require solutions of Riccati equations because solving such equations even once is a formidable task, let alone multiple times or in real time.

The most remarkable feature of backstepping designs is that they result in explicit formulae for the control gain (or in rapidly convergent symbolic or numerical iterative schemes for the computation of the control gains, which can be run online). The backstepping gain functions are explicit both in the spatial variables and in the physical parameters. It is the explicit dependence of the gain function on the physical parameters that is useful in adaptive control implementations, which we illustrate in this chapter. We show how one can compute an estimate of an unknown parameter online, plug it into the explicit gain formula derived earlier in the book (Chapter 4), and achieve closed-loop stability. This strategy is generally referred to as the *certainty equivalence* approach.

In this chapter we design certainty equivalence adaptive controllers with two types of identifiers: *passivity-based* identifiers and *swapping* identifiers.

The passivity-based method (often called the "observer-based" method) uses a copy of the plant to generate a model which is passive from the parameter estimation error as the model's input to the spatial inner product of the "observer error" and a "regressor" function as the model's output.

The swapping method (often called the "gradient" or "least squares" method) is the most common method of parameter estimation in adaptive control. Filters of the "regressor" and of the measured part of the plant are implemented to convert a dynamic parameterization of the problem (given by the plant's dynamic model) into a static parameterization, where standard gradient and least squares estimation techniques can be used. The swapping method, which employs "one filter per unknown parameter," has a higher dynamic order than the passivity-based method, which uses only one filter. However, the swapping approach can handle output-feedback problems and is more transparent due to the static parameterization.

Closed-loop stability of a system consisting of the plant, the identifier, and the controller is the central issue in adaptive control since the parameter estimates make the adaptive controller nonlinear even when the PDE plant is linear.

We will illustrate the ideas of the designs and the proofs in two benchmark examples.

The general theory of adaptive control of PDE systems based on the backstepping approach is beyond the scope of this introductory text and is the subject of a separate research monograph by the same authors [160].

13.1 State-Feedback Design with Passive Identifier

Consider a reaction-diffusion equation

$$u_t = u_{xx} + \lambda u , \tag{13.1}$$

$$u(0) = 0 , \tag{13.2}$$

where a constant parameter λ is *unknown*. From Chapter 3 we know that this system (with $u(1,t) = 0$) is unstable for any $\lambda > \pi^2$.

In Chapter 4 we designed the following stabilizing controller for the system (13.1)–(13.2):

$$u(1) = -\int_0^1 \lambda y \frac{I_1\left(\sqrt{\lambda(1-y^2)}\right)}{\sqrt{\lambda(1-y^2)}} u(y)\, dy . \tag{13.3}$$

We obviously cannot use this controller here since the parameter λ is unknown. Following the certainty equivalence principle, we need to design an identifier which will provide the estimate of λ.

13.1.1 Identifier

Let us denote the estimate of λ by $\hat{\lambda} = \hat{\lambda}(t)$ and introduce the following auxiliary system:

$$\hat{u}_t = \hat{u}_{xx} + \hat{\lambda} u + \gamma^2 (u - \hat{u}) \int_0^1 u^2(x)\, dx , \tag{13.4}$$

$$\hat{u}(0) = 0 , \tag{13.5}$$

$$\hat{u}(1) = u(1) . \tag{13.6}$$

Such an auxiliary system is often called an "observer," even though it is not used here for state estimation (the entire state u is available for measurement in our problem).

The purpose of this "observer" is not the estimation of the system state, since the full state $u(x,t)$ is measured. Its purpose is to help identify the unknown parameter, as we shall soon see.

We refer to the system (13.4)–(13.6) as a "passive identifier." This identifier employs a copy of the PDE plant and an additional nonlinear term. The term "passive identifier" is derived from the fact that an operator from the parameter estimation error $\tilde{\lambda} = \lambda - \hat{\lambda}$ to the inner product of u with $u - \hat{u}$ is strictly passive. The additional term in (13.4) acts as nonlinear damping, whose task is to slow down the adaptation.

Let us introduce the error signal

$$e = u - \hat{u}\,. \tag{13.7}$$

Using (13.1)–(13.2) and (13.4)–(13.6), we obtain the following PDE for $e(x,t)$:

$$e_t = e_{xx} + \tilde{\lambda} u - \gamma^2 e \|u\|^2\,, \tag{13.8}$$
$$e(0) = 0\,, \tag{13.9}$$
$$e(1) = 0\,. \tag{13.10}$$

We want to find out what stability properties are guaranteed by the identifier (13.4)–(13.6) for the signals e (estimation error) and

$$\tilde{\lambda} = \lambda - \hat{\lambda} \tag{13.11}$$

(parameter error). With Lyapunov function

$$V = \frac{1}{2}\int_0^1 e^2(x)\,dx + \frac{\tilde{\lambda}^2}{2\gamma}, \tag{13.12}$$

we get

$$\dot{V} = -\|e_x\|^2 - \gamma^2\|e\|^2\|u\|^2 + \tilde{\lambda}\int_0^1 e(x)u(x)\,dx - \frac{\tilde{\lambda}\dot{\hat{\lambda}}}{\gamma}. \tag{13.13}$$

With the choice of the update law

$$\dot{\hat{\lambda}} = \gamma \int_0^1 (u(x) - \hat{u}(x))u(x)\,dx\,, \tag{13.14}$$

the last two terms in (13.13) cancel out and we obtain

$$\dot{V} = -\|e_x\|^2 - \gamma^2\|e\|^2\|u\|^2\,, \tag{13.15}$$

which implies

$$V(t) \le V(0)\,. \tag{13.16}$$

By the definition of V, this means that $\tilde{\lambda}$ and $\|e\|$ are bounded functions of time.

Integrating (13.15) with respect to time from zero to infinity, we get

$$V(\infty) + \int_0^\infty \|e_x(t)\|^2\,dt + \gamma^2 \int_0^\infty \|e(t)\|^2\|u(t)\|^2\,dt \le V(0), \tag{13.17}$$

so that the spatial norms $\|e_x\|$ and $\|e\|\,\|u\|$ are square integrable functions of time. From the update law (13.14), we get

$$|\dot{\hat{\lambda}}| \le \gamma\|e\|\,\|u\|\,, \tag{13.18}$$

which shows that $\dot{\hat{\lambda}}$ is also square integrable in time; i.e., the identifier indirectly slows down the adaptation (without explicit normalization in the update law).

13.1.2 Target System

The next step is to investigate effects of the time-varying $\hat{\lambda}$ on the target system (which is a simple heat equation in the nonadaptive case).

For the unknown λ we transform the identifier rather than the plant:

$$\hat{w}(x) = \hat{u}(x) - \int_0^x \hat{k}(x,y)\hat{u}(y)\,dy\,, \tag{13.19}$$

$$\hat{k}(x,y) = -\hat{\lambda}\xi \frac{I_1\left(\sqrt{\hat{\lambda}(x^2-y^2)}\right)}{\sqrt{\hat{\lambda}(x^2-\xi^2)}}\,. \tag{13.20}$$

One can show that the above transformation maps (13.4)–(13.6) into the following target system:

$$\hat{w}_t = \hat{w}_{xx} + \dot{\hat{\lambda}} \int_0^x \frac{\xi}{2}\hat{w}(\xi)\,d\xi + (\hat{\lambda} + \gamma^2\|u\|^2)e_1\,, \tag{13.21}$$

$$\hat{w}(0) = 0\,, \tag{13.22}$$

$$\hat{w}(1) = 0\,, \tag{13.23}$$

where e_1 is the transformed estimation error

$$e_1(x) = e(x) - \int_0^x \hat{k}(x,\xi)e(y)\,dy\,. \tag{13.24}$$

We observe that, in comparison to the nonadaptive target system, two additional terms appear in (13.21), where one is proportional to $\dot{\hat{\lambda}}$ and the other is proportional to the estimation error e. The identifier guarantees that both of these terms are square integrable in time, which, loosely speaking, means that they decay to zero barring some occasional "spikes."

For further analysis we will also need the inverse transformation

$$\hat{u}(x) = \hat{w}(x) + \int_0^x \hat{l}(x,y)\hat{w}(y)\,dy\,, \tag{13.25}$$

$$\hat{l}(x,y) = -\hat{\lambda}y \frac{J_1\left(\sqrt{\hat{\lambda}(x^2-y^2)}\right)}{\sqrt{\hat{\lambda}(x^2-y^2)}}\,. \tag{13.26}$$

13.1.3 Boundedness of Signals

Before we proceed, let us state two useful results.

Lemma 13.1. *Suppose that the function $f(t)$ defined on $[0,\infty)$ satisfies the following conditions:*

- *$f(t) \geq 0$ for all $t \in [0,\infty)$;*

- *$f(t)$ is differentiable on $[0, \infty)$, and there exists a constant M such that*

$$f'(t) \le M \text{ for all } t \ge 0; \tag{13.27}$$

- $\int_0^\infty f(t)\,dt < \infty$.

Then

$$\lim_{t\to\infty} f(t) = 0. \tag{13.28}$$

This lemma, proved in [116], is an alternative (not a corollary) to "Barbalat's lemma," a standard tool in adaptive control.

Lemma 13.2 [97]. *Let v, l_1, and l_2 be real-valued functions of time defined on $[0, \infty)$, and let c be a positive constant. If l_1 and l_2 are nonnegative and integrable on $[0, \infty)$ and satisfy the differential inequality*

$$\dot v \le -cv + l_1(t)v + l_2(t), \quad v(0) \ge 0, \tag{13.29}$$

then v is bounded and integrable on $[0, \infty)$.

In Section 13.1.1 we proved that $\tilde\lambda$ is bounded. This implies that $\hat\lambda$ is also bounded; let us denote this bound by λ_0. The functions $\hat k(x, y)$ and $\hat l(x, y)$ are bounded and twice continuously differentiable with respect to x and y; therefore there exist constants M_1, M_2, M_3 such that

$$\|e_1\| \le M_1\|e\|, \tag{13.30}$$

$$\|u\| \le \|\hat u\| + \|e\| \le M_2\|\hat w\| + \|e\|, \tag{13.31}$$

$$\|u_x\| \le \|\hat u_x\| + \|e_x\| \le M_3\|\hat w_x\| + \|e_x\|. \tag{13.32}$$

Using Young's, Cauchy–Schwarz, and Poincaré inequalities, we estimate

$$\begin{aligned}
\frac{1}{2}\frac{d}{dt}\|\hat w\|^2 &= -\int_0^1 \hat w_x^2\,dx + \dot{\hat\lambda}\int_0^1 \hat w(x)\int_0^x \frac{\xi}{2}\hat w(\xi)\,d\xi\,dx + (\hat\lambda + \gamma^2\|u\|^2)\int_0^1 e_1\hat w\,dx \\
&\le -\|\hat w_x\|^2 + \frac{|\dot{\hat\lambda}|}{2}\|\hat w\|^2 + M_1\lambda_0\|\hat w\|\|e\| + \gamma^2 M_1\|u\|(M_2\|\hat w\| + \|e\|)\|\hat w\|\|e\| \\
&\le -\frac{1}{4}\|\hat w\|^2 + \frac{1}{16}\|\hat w\|^2 + |\dot{\hat\lambda}|^2\|\hat w\|^2 + \frac{1}{16}\|\hat w\|^2 \\
&\quad + 4M_1^2\lambda_0^2\|e\|^2 + \frac{1}{16}\|\hat w\|^2 + 8\gamma^4 M_1^2 M_2^2\|u\|^2\|e\|^2\|\hat w\|^2 + \frac{\|e\|^2}{16M_2^2} \\
&\le -\frac{1}{16}\|\hat w\|^2 + l_1\|\hat w\|^2 + l_2,
\end{aligned} \tag{13.33}$$

where l_1, l_2 are some integrable functions of time on $[0, \infty)$. The last inequality follows from the square integrability of $\dot{\hat\lambda}$, $\|u\|\|e\|$, and $\|e\|$. Using Lemma 13.2 we get the boundedness and integrability of $\|\hat{w}\|^2$ (or square integrability of $\|\hat{w}\|$). From (13.30) we get boundedness and square integrability of $\|u\|$ and $\|\hat{u}\|$, and (13.14) implies that $\dot{\hat{\lambda}}$ is bounded.

In order to get pointwise in x boundedness, we need to show the boundedness of $\|\hat{w}_x\|$ and $\|e_x\|$:

$$
\begin{aligned}
\frac{1}{2}\frac{d}{dt}\int_0^1 \hat{w}_x^2\,dx &= \int_0^1 \hat{w}_x\hat{w}_{xt}\,dx = -\int_0^1 \hat{w}_{xx}\hat{w}_t\,dx \\
&= -\int_0^1 \hat{w}_{xx}^2\,dx - \frac{\dot{\hat{\lambda}}}{2}\int_0^1 \hat{w}_{xx}\int_0^x \xi w(\xi)\,d\xi\,dx \\
&\quad - (\hat{\lambda} + \gamma^2\|u\|^2)\int_0^1 e_1\hat{w}_{xx}\,dx \\
&\le -\frac{1}{8}\|\hat{w}_x\|^2 + \frac{|\dot{\hat{\lambda}}|^2\|\hat{w}\|^2}{4} + (\lambda_0 + \gamma^2\|u\|^2)^2 M_1\|e\|^2\,, \qquad (13.34)
\end{aligned}
$$

$$
\begin{aligned}
\frac{1}{2}\frac{d}{dt}\int_0^1 e_x^2\,dx &\le -\|e_{xx}\|^2 + |\tilde{\lambda}|\|e_{xx}\|\|u\| - \gamma^2\|e_x\|^2\|u\|^2 \\
&\le -\frac{1}{8}\|e_x\|^2 + \frac{1}{2}|\tilde{\lambda}|^2\|u\|^2\,. \qquad (13.35)
\end{aligned}
$$

Since the right-hand sides of (13.34) and (13.35) are integrable, using Lemma 13.2 we get boundedness and square integrability of $\|\hat{w}_x\|$ and $\|e_x\|$. From (13.25) we get boundedness and square integrability of $\|\hat{u}_x\|$ and, as a consequence, of $\|u_x\|$. By the Agmon inequality,

$$
\max_{x\in[0,1]} |u(x,t)|^2 \le 2\|u\|\|u_x\| \le \infty \qquad (13.36)
$$

so that u (and, similarly, $\hat{u}$) is bounded for all $x \in [0, 1]$.

13.1.4 Regulation

To show the regulation of u to zero, note that

$$
\frac{1}{2}\frac{d}{dt}\|e\|^2 \le -\|e_x\|^2 + |\tilde{\lambda}|\|e\|\|u\| < \infty\,. \qquad (13.37)
$$

The boundedness of $(d/dt)\|w\|^2$ follows from (13.33). Using Lemma 13.1 we get

$$
\|\hat{w}\| \to 0,\ \|e\| \to 0 \text{ as } t \to \infty.
$$

It follows from (13.25) that

$$
\|\hat{u}\| \to 0,
$$

and therefore

$$
\|u\| \to 0.
$$

Using the Agmon inequality and the fact that $\|u_x\|$ is bounded, we get the regulation of $u(x,t)$ to zero for all $x \in [0, 1]$:

$$
\lim_{t\to\infty}\max_{x\in[0,1]} |u(x,t)| \le \lim_{t\to\infty}(2\|u\|\|u_x\|)^{1/2} = 0\,. \qquad (13.38)
$$

The summary of the passivity-based design is presented in Table 13.1.

Table 13.1. *Summary of the adaptive design with passive identifier.*

Plant:

$$u_t = u_{xx} + \lambda u\,, \tag{13.39}$$
$$u(0) = 0\,. \tag{13.40}$$

Identifier:

$$\hat{u}_t = \hat{u}_{xx} + \hat{\lambda} u + \gamma^2 (u - \hat{u}) \int_0^1 u^2(x)\,dx\,, \tag{13.41}$$
$$\hat{u}(0) = 0\,, \tag{13.42}$$
$$\hat{u}(1) = u(1)\,, \tag{13.43}$$

$$\dot{\hat{\lambda}} = \gamma \int_0^1 (u(x) - \hat{u}(x)) u(x)\,dx\,. \tag{13.44}$$

Controller:

$$u(1) = -\int_0^1 \hat{\lambda} y \frac{I_1\left(\sqrt{\hat{\lambda}(1-y^2)}\right)}{\sqrt{\hat{\lambda}(1-y^2)}} \hat{u}(y)\,dy\,. \tag{13.45}$$

13.2 Output-Feedback Design with Swapping Identifier

As we had indicated in the introduction to this chapter, we give examples only of adaptive designs for PDEs; a wider, more general presentation is beyond the scope of this book. In Section 13.1 we provided an example of a state feedback adaptive design. In this section we provide an example of an output-feedback adaptive design.

Consider the plant

$$u_t = u_{xx} + g u(0)\,, \tag{13.46}$$
$$u_x(0) = 0 \tag{13.47}$$

with only $u(0)$ measured and $u(1)$ actuated.

This system is motivated by a model of thermal instability in solid propellant rockets [12]. The open-loop system (with $u(1) = 0$) is unstable if and only if $g > 2$.

The plant can be written in the frequency domain as a transfer function from input $u(1)$ to output $u(0)$:

$$u(0, s) = \frac{s}{(s-g)\cosh\sqrt{s} + g} u(1, s)\,. \tag{13.48}$$

We can see that it has no zeros (at $s = 0$ the transfer function is $2/(2-g)$) and has infinitely many poles, one of which is unstable and approximately equal to g as $g \to +\infty$. Hence, this is an infinite relative degree system.

13.2.1 Identifier

We employ two filters: the state filter

$$v_t = v_{xx} + u(0)\,, \tag{13.49}$$
$$v_x(0) = 0\,, \tag{13.50}$$
$$v(1) = 0 \tag{13.51}$$

and the input filter

$$\eta_t = \eta_{xx}\,, \tag{13.52}$$
$$\eta_x(0) = 0\,, \tag{13.53}$$
$$\eta(1) = u(1)\,. \tag{13.54}$$

The role of filters is to convert the dynamic parametrization (the plant) into a static parametrization, which is obtained as follows.

We introduce the "estimation" error

$$e(x) = u(x) - gv(x) - \eta(x)\,, \tag{13.55}$$

which is exponentially stable:

$$e_t = e_{xx}\,, \tag{13.56}$$
$$e_x(0) = 0\,, \tag{13.57}$$
$$e(1) = 0\,. \tag{13.58}$$

Then we set $x = 0$ in (13.55) (since only $u(0)$ is measured) and take the resulting equation as a parametric model:

$$e(0) = u(0) - gv(0) - \eta(0)\,. \tag{13.59}$$

Let us denote the estimate of g by $\hat{g}$.

We choose the standard gradient update law with normalization

$$\dot{\hat{g}} = \gamma \frac{\hat{e}(0)v(0)}{1 + v^2(0)}\,, \tag{13.60}$$

where $\hat{e}(0)$ is a "prediction error":

$$\hat{e}(0) = u(0) - \hat{g}v(0) - \eta(0)\,. \tag{13.61}$$

Using the Lyapunov function

$$V = \frac{1}{2}\int_0^1 e^2\,dx + \frac{1}{2\gamma}\tilde{g}^2\,, \tag{13.62}$$

where $\tilde{g} = g - \hat{g}$ is the parameter error, we get

$$\begin{aligned}\dot{V} &= -\int_0^1 e_x^2\,dx - \frac{\tilde{g}\hat{e}(0)v(0)}{1+v^2(0)} \\ &\le -\int_0^1 e_x^2\,dx - \frac{\hat{e}^2(0)}{1+v^2(0)} + \frac{e(0)\hat{e}(0)}{1+v^2(0)} \\ &\le -\|e_x\|^2 - \frac{\hat{e}^2(0)}{1+v^2(0)} + \frac{\|e_x\|\,|\hat{e}(0)|}{\sqrt{1+v^2(0)}} \\ &\le -\frac{1}{2}\|e_x\|^2 - \frac{1}{2}\frac{\hat{e}^2(0)}{1+v^2(0)}\,. \end{aligned} \tag{13.63}$$

Therefore,

$$V(t) \le V(0)\,, \tag{13.64}$$

and from the definition of V we get boundedness of $\|e\|$ and $\tilde{g}$. Integrating (13.63) in time from zero to infinity, we can see that $\frac{\hat{e}(0)}{\sqrt{1+v^2(0)}}$ is a square integrable function of time on $[0,\infty)$. Since

$$\frac{\hat{e}(0)}{\sqrt{1+v^2(0)}} = \frac{e(0)}{\sqrt{1+v^2(0)}} + \tilde{g}\frac{v(0)}{\sqrt{1+v^2(0)}}\,, \tag{13.65}$$

$\frac{\hat{e}(0)}{\sqrt{1+v^2(0)}}$ is also bounded. Writing the update law (13.60) as

$$\dot{\hat{g}} = \gamma\frac{\hat{e}(0)}{\sqrt{1+v^2(0)}}\frac{v(0)}{\sqrt{1+v^2(0)}}\,, \tag{13.66}$$

we get boundedness and square integrability of $\dot{\hat{g}}$.

Remark 13.3. Note that the passive identifier did not guarantee boundedness of the time derivative of the parameter estimate. However, thanks to the normalization in the update law, the swapping identifier does guarantee that.

13.2.2 Controller

In Chapter 9 we designed the following controller for (13.46), (13.47) when g is known:

$$u(1) = -\int_0^1 \sqrt{g}\sinh\sqrt{g}(1-y)u(y)\,dy\,. \tag{13.67}$$

Suppose now that g is unknown. According to the certainty equivalence principle, we modify (13.67) in two ways: we replace g with its estimate $\hat{g}$ and replace the unmeasured state u with its estimate $\hat{g}v + \eta$.

The resulting controller is

$$u(1,t) = \int_0^1 \hat{k}(1,y)(\hat{g}v(y) + \eta(y))\,dy\,, \tag{13.68}$$

$$\hat{k}(x,y) = \begin{cases} -\sqrt{\hat{g}}\sinh\sqrt{\hat{g}}(x-y), & \hat{g}\ge 0, \\ \sqrt{-\hat{g}}\sin\sqrt{-\hat{g}}(x-y), & \hat{g} < 0, \end{cases} \tag{13.69}$$

where the control kernel is written in a form which acknowledges that $\hat{g}$ may become negative during the transient.

13.2.3 Target System

The backstepping transformation is

$$\hat{w}(x) = \hat{g}v(x) + \eta(x) - \int_0^x \hat{k}(x,y)(\hat{g}v(y) + \eta(y))\,dy\,, \tag{13.70}$$

where $\hat{k}(x,y)$ is given by (13.69). One can show that it maps (13.46), (13.47), along with (13.68), into the following system:

$$\hat{w}_t = \hat{w}_{xx} + \beta(x)\hat{e}(0) + \dot{\hat{g}}\left[v + \int_0^x \alpha(x-y)\left(\hat{g}v(y) + \hat{w}(y)\right)dy\right], \tag{13.71}$$

$$\hat{w}_x(0) = 0\,, \tag{13.72}$$

$$\hat{w}(1) = 0\,, \tag{13.73}$$

where

$$\alpha(x) = -\frac{1}{\hat{g}}\hat{k}(x,0)\,, \tag{13.74}$$

$$\beta(x) = \hat{k}_\xi(x,0) = \begin{cases} \hat{g}\cosh\sqrt{\hat{g}}x, & \hat{g} \geq 0\,, \\ \hat{g}\cos\sqrt{-\hat{g}}x, & \hat{g} < 0\,. \end{cases} \tag{13.75}$$

Note that, when we compare this to the nonadaptive target system (heat equation), we see that we have two extra terms: one is proportional to $\hat{e}(0)$ and the other is proportional to $\dot{\hat{g}}$. The identifier guarantees that these terms decay (with occasional "spikes"); however, we have an extra state in the equation: the filter v. Therefore we rewrite the equation for v as

$$v_t = v_{xx} + \hat{w}(0) + \hat{e}(0)\,, \tag{13.76}$$

$$v_x(0) = 0\,, \tag{13.77}$$

$$v(1) = 0 \tag{13.78}$$

and analyze the interconnection of two systems $\hat{w}$, v driven by the external signal $\hat{e}(0)$.

13.2.4 Boundedness of Signals

Let us denote the bounds on $\hat{g}$, α, and β by g_0, α_0, and β_0, correspondingly.

In order to establish boundedness of $\|w\|$ and $\|v\|$, consider the Lyapunov function

$$V_v = \frac{1}{2}\int_0^1 v^2(x)\,dx + \frac{1}{2}\int_0^1 v_x^2(x)\,dx\,. \tag{13.79}$$

Using Young's and Poincaré inequalities, we have

$$\begin{aligned}
\dot{V}_v &= -\int_0^1 v_x^2\,dx + (\hat{w}(0)+\hat{e}(0))\int_0^1 v\,dx \\
&\quad - \int_0^1 v_{xx}^2\,dx - (\hat{w}(0)+\hat{e}(0))\int_0^1 v_{xx}\,dx \\
&\le -\|v_x\|^2 + \frac{1}{8}\|v\|^2 + 4\frac{\hat{e}^2(0)}{1+v^2(0)}(1+\|v_x\|^2) \\
&\quad + 4\|\hat{w}_x\|^2 - \|v_{xx}\|^2 + \frac{1}{2}\|v_{xx}\|^2 + \|\hat{w}_x\|^2 \\
&\quad + \frac{\hat{e}^2(0)}{1+v^2(0)}(1+\|v_x\|^2) \\
&\le -\frac{1}{2}\|v_x\|^2 - \frac{1}{2}\|v_{xx}\|^2 + 5\|\hat{w}_x\|^2 + l_1\|v_x\|^2 + l_2,
\end{aligned} \tag{13.80}$$

where l_1, l_2 are some integrable functions of time (due to the properties guaranteed by the identifier). Using the following Lyapunov function for the $\hat{w}$-system:

$$V_{\hat{w}} = \frac{1}{2}\int_0^1 \hat{w}^2(x)\,dx\,, \tag{13.81}$$

we get

$$\begin{aligned}
\dot{V}_{\hat{w}} &= -\int_0^1 \hat{w}_x^2\,dx + \hat{e}(0)\int_0^1 \beta\hat{w}\,dx + \dot{\hat{g}}\int_0^1 \hat{w}v\,dx \\
&\quad + \dot{\hat{g}}\int_0^1 \hat{w}(x)\int_0^x \alpha(x-y)(\hat{g}v(y)+\hat{w}(y))\,dy\,dx \\
&\le -\|\hat{w}_x\|^2 + \frac{c_1}{2}\|\hat{w}\|^2 + \frac{\beta_0^2}{2c_1}\frac{\hat{e}^2(0)}{1+v^2(0)}(1+\|v_x\|^2) \\
&\quad + \frac{|\dot{\hat{g}}|^2(1+\alpha_0 g_0)^2}{2c_1}\|v\|^2 + c_1\|\hat{w}\|^2 + \frac{|\dot{\hat{g}}|^2\alpha_0^2}{2c_1}\|\hat{w}\|^2 \\
&\le -(1-6c_1)\|\hat{w}_x\|^2 + l_3\|\hat{w}\|^2 + l_4\|v_x\|^2 + l_5\,,
\end{aligned} \tag{13.82}$$

where l_3, l_4, and l_5 are integrable functions of time. Choosing $c_1 = 1/24$ and using the Lyapunov function

$$V = V_{\hat{w}} + \frac{1}{20}V_v\,,$$

we get

$$\begin{aligned}
\dot{V} &\le -\frac{1}{2}\|\hat{w}_x\|^2 - \frac{1}{40}\|v_x\|^2 - \frac{1}{40}\|v_{xx}\|^2 \\
&\quad + l_3\|\hat{w}\|^2 + \left(l_4 + \frac{l_1}{20}\right)\|v_x\|^2 + l_5 + \frac{l_2}{20} \\
&\le -\frac{1}{4}V + l_6 V + l_7\,,
\end{aligned} \tag{13.83}$$

where l_6 and l_7 are integrable. By Lemma 13.2 we proved boundedness and square integrability of $\|\hat{w}\|$,$\|v\|$, and $\|v_x\|$. Using these properties we get

$$\begin{aligned}\frac{1}{2}\frac{d}{dt}\|\hat{w}_x\|^2 &\le -\|\hat{w}_{xx}\|^2 + \beta_0|\hat{e}(0)|\|\hat{w}_{xx}\| + |\dot{\hat{g}}|\|\hat{w}_{xx}\|((1+\alpha_0 g_0)\|v\| + \alpha_0\|\hat{w}\|)\\ &\le -\frac{1}{8}\|\hat{w}_x\|^2 + l_8,\end{aligned} \tag{13.84}$$

where l_8 is integrable, and thus $\|\hat{w}_x\|$ is bounded and square integrable. By the Agmon inequality we get boundedness of $w(x,t)$ and $v(x,t)$ for all $x \in [0,1]$.

13.2.5 Regulation

Using the fact that $\|v_x\|$, $\|\hat{w}_x\|$ are bounded, one can easily show that

$$\left|\frac{d}{dt}(\|v\|^2 + \|\hat{w}\|^2)\right| < \infty. \tag{13.85}$$

By Lemma 13.1,

$$\|\hat{w}\| \to 0, \; \|v\| \to 0.$$

From the transformation, inverse to (13.70),

$$\hat{g}v(x) + \eta(x) = \hat{w}(x) - \hat{g}\int_0^x (x-y)\hat{w}(y)\,dy\,, \tag{13.86}$$

we have that

$$\|\eta\| \to 0$$

and that $\|\eta_x\|$ is bounded. Using (13.55) we get

$$\|u\| \to 0$$

and the boundedness of $\|u_x\|$. Finally, using the Agmon inequality we get

$$\lim_{t\to\infty}\max_{x\in[0,1]}|u(x,t)| \le \lim_{t\to\infty}(2\|u\|\|u_x\|)^{1/2} = 0\,. \tag{13.87}$$

The summary of the design with swapping identifier is presented in Table 13.2.

In Figure 13.1 the simulation results for the scheme with swapping identifier are presented. We can see that the plant is stabilized and the parameter estimate comes close to the true value $g = 5$ (sufficiently close so that the controller is stabilizing, but not converging to the true value because of the absence of persistency of excitation).

Table 13.2. *Summary of the adaptive design with swapping identifier.*

Plant:

$$u_t = u_{xx} + gu(0)\,, \tag{13.88}$$
$$u_x(0) = 0\,. \tag{13.89}$$

Filters:

$$v_t = v_{xx} + u(0)\,, \tag{13.90}$$
$$v_x(0) = 0\,, \tag{13.91}$$
$$v(1) = 0\,, \tag{13.92}$$

$$\eta_t = \eta_{xx}\,, \tag{13.93}$$
$$\eta_x(0) = 0\,, \tag{13.94}$$
$$\eta(1) = u(1)\,. \tag{13.95}$$

Update law:

$$\dot{\hat{g}} = \gamma \frac{(u(0) - \hat{g}v(0) - \eta(0))v(0)}{1 + v^2(0)}\,. \tag{13.96}$$

Controller:

$$u(1) = \int_0^1 \hat{k}(1, y)(\hat{g}v(y) + \eta(y))\,dy\,, \tag{13.97}$$

$$\hat{k}(1, y) = \begin{cases} -\sqrt{\hat{g}}\sinh\sqrt{\hat{g}}(1 - y), & \hat{g} \geq 0\,, \\ \sqrt{-\hat{g}}\sin\sqrt{-\hat{g}}(1 - y), & \hat{g} < 0\,. \end{cases} \tag{13.98}$$

13.3 Notes and References

Early work on adaptive control of infinite-dimensional systems focused on plants stabilizable by nonidentifier-based high gain feedback [121] under a relative degree one assumption.

State-feedback model reference adaptive control (MRAC) was extended to PDEs in [73, 25, 161, 135, 23], but not for the case of boundary control. Efforts in [46, 184] made use of positive realness assumptions: where relative degree one is implicit.

Stochastic adaptive linear quadratic regulator (LQR) with least-squares parameter estimation and state feedback was pursued in [51].

Adaptive control of nonlinear PDEs was studied in [116, 88]. Adaptive controllers for nonlinear systems on lattices were designed in [79].

An experimentally validated adaptive boundary controller for a flexible beam was presented in [144].

An overview and systematization of the adaptive backstepping techniques for PDEs are presented in [94]. A Lyapunov approach to identifier design is considered in [98].

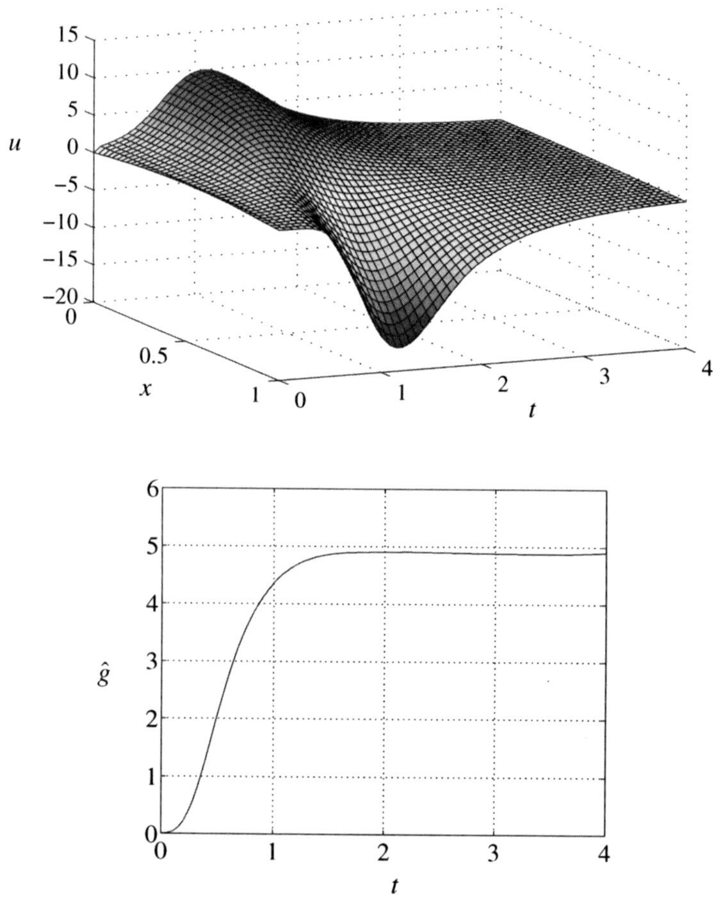

Figure 13.1. *The closed-loop simulation of the adaptive scheme with swapping identifier. Top: The closed-loop response. Bottom: Evolution of the parameter estimate.*

Exercises

13.1. Consider the plant

$$u_t = u_{xx} + bu_x + \lambda u\,,$$
$$u(0) = 0\,,$$

where b and λ are *unknown* constants and $u(1)$ is the input. Following the certainty equivalence principle, design the adaptive scheme with passive identifier (the nominal controller for this plant was designed in Chapter 4). Use the identifier

$$\hat{u}_t = \hat{u}_{xx} + \hat{b}u_x + \hat{\lambda}u + \gamma^2(u - \hat{u}) \int_0^1 u_x^2(x)\,dx\,,$$
$$\hat{u}(0) = 0\,,$$
$$\hat{u}(1) = u(1)$$

and the Lyapunov function

$$V = \frac{1}{2}\int_0^1 e^2(x)\,dx + \frac{1}{2\gamma}(\tilde{\lambda}^2 + \tilde{b}^2).$$

Show that the update laws

$$\dot{\hat{\lambda}} = \gamma \int_0^1 (u(x) - \hat{u}(x))u(x)\,dx\,,$$
$$\dot{\hat{b}} = \gamma \int_0^1 (u(x) - \hat{u}(x))u_x(x)\,dx$$

guarantee that estimation error $\|e\|$ and parameter errors $\tilde{\lambda}$ and $\tilde{b}$ are bounded and that $\|e_x\|$, $\|e\|\,\|u_x\|$, $\dot{\hat{\lambda}}$, and $\dot{\hat{b}}$ are square integrable functions of time.

13.2. Consider the plant

$$u_t = u_{xx} + g u_x(0)\,,$$
$$u(0) = 0\,,$$

where g is an unknown parameter, $u_x(0)$ is measured; and $u_x(1)$ is actuated. Design the output-feedback adaptive controller for this plant. Follow these steps:

(1) Derive the nominal control law for the case when g is known. Use the target system

$$w_t = w_{xx}\,,$$
$$w(0) = 0\,,$$
$$w_x(1) = 0$$

and show that the gain kernel PDE is

$$k_{xx}(x, y) = k_{yy}(x, y)\,,$$
$$k(x, 0) = -g + g\int_0^x k(x, y)\,dy\,,$$
$$k(x, x) = -g\,.$$

To solve this PDE, set

$$k(x, y) = \phi(x - y)\,, \tag{13.99}$$

write down the integral equation for $\phi(x)$, apply the Laplace transform, solve for $\phi(s)$, and apply the inverse Laplace transform. You should get

$$k(x, y) = -g e^{g(x-y)}\,.$$

(2) Introduce the filters

$$
\begin{aligned}
v_t &= v_{xx} + u_x(0)\,, \\
v(0) &= 0\,, \\
v_x(1) &= 0\,,
\end{aligned}
$$

$$
\begin{aligned}
\eta_t &= \eta_{xx}\,, \\
\eta(0) &= 0\,, \\
\eta_x(1) &= u_x(1)
\end{aligned}
$$

and the "estimation" error

$$
e(x) = u(x) - g v(x) - \eta(x)\,.
$$

Show that this error satisfies an exponentially stable PDE.

(3) Use the Lyapunov function

$$
V = \frac{1}{2}\int_0^1 e_x^2\,dx + \frac{1}{2\gamma}\tilde{g}^2
$$

and follow (13.63) to show that

$$
\dot{V} = -\frac{1}{2}\int_0^1 e_{xx}^2\,dx - \frac{\hat{e}_x(0)^2}{1+v_x(0)^2}\,.
$$

(You will need to use the inequality $|e_x(0)| \leq \|e_{xx}\|$.) This proves that $\|e_x\|$ and $\tilde{g}$ are bounded.

Show that the update law

$$
\dot{\hat{g}} = \gamma \frac{\hat{e}_x(0) v_x(0)}{1+v_x(0)^2}
$$

guarantees boundedness and square integrability of $\dot{\hat{g}}$ (use system (13.65), (13.66) as an example).

(4) Write down the certainty equivalence controller.

Chapter 14

Towards Nonlinear PDEs

Even though the bulk of this book is dedicated to linear PDEs (except for Chapter 13 on adaptive control, where the designs are nonlinear even though they are made for linear PDEs), our real motivation for studying the backstepping methods comes from their potential for nonlinear PDEs. To put it plainly we ask, what method would one hope to be successful for nonlinear PDEs if not the method that has been the most effective from among the methods for ODEs?

The field of nonlinear control for PDEs is still in its infancy. Its growth to adulthood and maturity promises to be extremely challenging. The problems that have so far proved tractable have been problems in which the nonlinearity appears in the PDE in such a way that it does not affect stability (relative to a simple choice of a Lyapunov function). This is the case, for example, with [93] for the viscous Burgers equation, and various extensions of this result for the Burgers equation [9, 114, 116]; the Korteweg–de Vries equation [117, 10]; the Kuramoto–Sivashinsky equation [115]; and the Navier–Stokes equations [13, 8, 1]. Another notable result, which in fact involves one step of backstepping on a system consisting of a nonlinear PDE and two ODEs, was the result by Banaszuk, Hauksson, and Mezić [16] on stabilization of an infinite-dimensional Moore–Greitzer model of compressor rotating stall instability. Results such as the ones listed here, most of which involve nonlinear parabolic PDEs, exist also for hyperbolic and hyperbolic-like PDEs (nonlinear string, plate, etc. models) and are also characterized by nonhostile nonlinearities and the fact that simple Lyapunov functions can be used.

Problems where nonlinearities in PDEs do not need to be counteracted are in a sense easier (such as control design problems) than classical nonlinear control problems for ODEs [97]. The problems that are of the most practical interest, and also are the most mathematically challenging, are problems with boundary control and with in-domain nonlinearities of harmful type, where not only do complex nonlinear infinite-dimensional feedback laws need to be synthesized, but also complex Lyapunov functions need to be found. Currently, only a single such result exists, which is boundary control for parabolic PDEs with Volterre nonlinearities [175, 176, 177]. This result is far too complex to have a place in an introductory text; however, we will present a condensed version of this result, as it is in a way the punchline of this book, and because it is a template for the kind of nonlinear PDE problems that should be pursued in the field.

14.1 The Nonlinear Optimal Control Alternative

Even our highly simplified presentation of the design from [175, 176, 177] will reveal the daunting complexity of the design, which is linked to the complexity of the problem of stabilizing a nonlinear PDE with in-domain destabilizing nonlinearities using only boundary control. Before we begin, we want to discuss one possible alternative to our approach—a design based on nonlinear optimal control; to be precise, infinite-horizon, Hamilton–Jacobi–Bellman (HJB) PDE-based nonlinear optimal control.

To develop an HJB-based feedback law, the simplest approach would entail (semi-) discretizing the nonlinear PDE plant (in space) and developing an optimal feedback law for the resulting, ODE system. For example, consider the simplest nonlinear PDE boundary control problem worth studying as a boundary control problem with a possibly destabilizing in-domain nonlinearity, that is, a reaction-diffusion equation

$$u_t = u_{xx} + f(u)\,, \tag{14.1}$$

where $f(u)$ is a continuous function, for instance, $f(u) = u^2$. Let us consider a semi-discretization in x, which results in a set of ODEs. For instance, one would typically expect to need at least 10 grid points for a reasonable spatial discretization of a reaction-diffusion system, so the resulting ODE system would be evolving in $\mathbb{R}^{10}$.

To solve the HJB PDE for the resulting ODE control system, one would have to grid the state space of the ODE (or one can think of that as "quantization" of the state space). It would be reasonable to assume that at least 30 quantization points of the state space are needed (furthermore, one may employ a log/exp quantization of the state variables to even out the resolution of the state space for both large and small values of the state). Hence, we would have 10 ODEs, with each one being approximated to a 30-point grid, as shown in Figure 14.1.

The solution to the HJB equation would have to be stored in a 10D array, where, for each point on the state grid in the 10D state space, the value of the solution of the HJB equation is stored. *Provided the HJB equation can actually be solved* (which is a highly nontrivial problem), one needs 30^{10} bytes of memory, which is approximately 1 petabyte (PB) of storage space, to *store* the feedback law.

It is clear that 1 PB is a pretty large amount of disk space, but how large?

One can get a sense for the size of 1 PB of memory by noting that the entire disk space of Google™ in 2007 was 2 PBs.

That a coarsely quantized 10th-order nonlinear ODE system would require half of Google's disk space just to store an optimal feedback law is an extremely discouraging observation. Such an observation has been known for decades as the *curse of dimensionality* (often mentioned in the context of the "dynamic programming" approach to nonlinear control).

This dramatic observation carries two clear messages. The first is that one may want to give up *optimality* when pursuing control design for nonlinear PDEs; the second is that the format one pursues for the nonlinear feedback laws should not be a "precomputed feedback

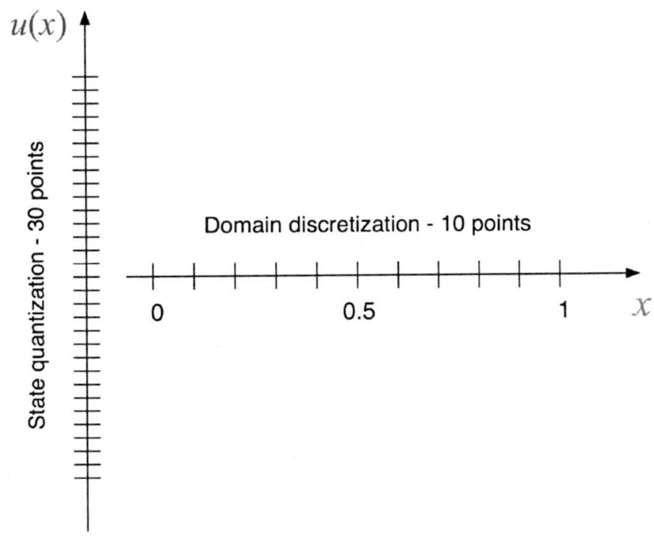

Figure 14.1. *Discretizing and quantizing the state space of the reaction-diffusion PDE* (14.1) *for the purpose of applying infinite-horizon HJB PDE-based nonlinear optimal control design.*

map" but a format where only the "gains" of a particular nonlinear feedback formula are precomputed and the actual control value is computed online, based on the measurements of the state.

The approach that we will consider in what follows results in feedback laws that are infinite-dimensional but in which the nonlinear feedback operators are parameterized in a polynomial manner using Volterra series in space. Hence, only the gain kernels (of the Volterra series) need to be precomputed and stored. For the particular example (14.1) that we will consider further in the chapter, the gain storage requirement is only about 1.5 (KBs), which is twelve orders of magnitude less than the 1 PB requirement mentioned above.

Our feedback laws won't be simple; quite on the contrary, however, their complexity will be commensurate with the plant rather than exceeding it by a large margin.

14.2 Feedback Linearization for a Nonlinear PDE: Transformation in Two Stages

Feedback linearization ideas have been considered for PDEs, using discretization in space or various forms of ODE model reduction. None have, however, been proved to be convergent as the discretization step goes to zero or as the reduced ODE model approaches the PDE limit.

In this chapter we present a design approach that involves no discretization or model reduction and which results in nonlinear feedback laws based on the backstepping approach developed in the previous chapters.

The design approach that we present here is applicable to a very general class of plants, but it is remarkably involved in the general case. For clarity of presentation, in this chapter we aim to give the reader only a flavor of the methodology for nonlinear control design for PDEs, and we do this through the following example:

$$v_t = v_{xx} + v^2, \qquad 0 < x < 1, \tag{14.2}$$

$$v(0,t) = 0, \tag{14.3}$$

$$v_x(1,t) = \text{boundary control.} \tag{14.4}$$

This system suffers from finite escape time instability in the absence of control.

In order to pursue a feedback linearizing design for this PDE, it helps to recall the basic principles of feedback linearization for ODEs. *Full-state* feedback linearization for ODEs requires that a *full relative degree output* (the output whose relative degree is equal to the order of the plant) be guessed correctly. (The Lie bracket tests allow us to establish whether such an output *exists* but do not show *how* such an output can be found.)

For the present example, the *full relative degree output* is

$$v_x(0,t). \tag{14.5}$$

This choice is motivated by the Dirichlet boundary condition at $x = 0$ (thus we select a Neumann output) and by the fact that the control enters at the opposite boundary, $x = 1$.

The feedback linearizing design then proceeds in two stages; namely, the linearizing transformation is a composition of two transformations.

Transformation 1. The first transformation is very simple (one should view it as a "pre-transformation") and is given by

$$u(x,t) = v_x(x,t). \tag{14.6}$$

This transformation converts the system (14.2)–(14.4) into

$$u_t(x,t) = u_{xx}(x,t) + 2u(x,t)\int_0^x u(\xi,t)d\xi, \tag{14.7}$$

$$u_x(0,t) = 0, \tag{14.8}$$

$$u(1,t) = \text{boundary control.} \tag{14.9}$$

The transformation $u = v_x$ is appealing from the regularity point of view, as the stabilization of u in L_2 will automatically yield both H_1 and L_2 stability of v (due to the boundary condition $v(0,t) = 0$). However, the usefulness of the transformation is not obvious yet, as the simple function $v(x,t)^2$ has been replaced by a more complicated functional $2u(x,t)\int_0^x u(\xi,t)d\xi$.

Transformation 2. The heart of the feedback linearizing transformation is a backstepping transformation. Guided by the Volterra form of the backstepping transformation in the linear case, we pursue the nonlinear transformation in the form of a Volterra series in the spatial variable x, namely,

$$w(x) = u(x) - \int_0^x u(\xi_1) \int_0^{\xi_1} u(\xi_2) k_2(x, \xi_1, \xi_2) d\xi_2 d\xi_1$$
$$- \int_0^x u(\xi_1) \int_0^{\xi_1} u(\xi_2) \int_0^{\xi_2} u(\xi_3) k_3(x, \xi_1, \xi_2, \xi_3) d\xi_3 d\xi_2 d\xi_1$$
$$- \cdots \quad \text{(terms of order 4 and higher).} \tag{14.10}$$

The target system for this transformation is the heat equation,

$$w_t = w_{xx}, \tag{14.11}$$
$$w_x(0, t) = 0, \tag{14.12}$$
$$w(1, t) = 0, \tag{14.13}$$

where the quadratic (destabilizing) reaction term v^2 has been eliminated.

The boundary condition (14.13) is achieved with the control law

$$u(1) = \int_0^1 u(\xi_1) \int_0^{\xi_1} u(\xi_2) k_2(1, \xi_1, \xi_2) d\xi_2 d\xi_1$$
$$+ \int_0^1 u(\xi_1) \int_0^{\xi_1} u(\xi_2) \int_0^{\xi_2} u(\xi_3) k_3(1, \xi_1, \xi_2, \xi_3) d\xi_3 d\xi_2 d\xi_1 + \cdots . \tag{14.14}$$

The reader should note that this feedback law has no linear component because the linearized plant is the heat equation, which is exponentially stable, and thus one need not alter the linear part of the closed-loop system. Hence, we use control with only nonlinear terms (powers of two and higher), whose sole purpose is to prevent finite escape.

With powers of u going to infinity in (14.10) and (14.14), and having infinitely many terms in the Volterra series, a crucial question arises—what are the chances that such a control law would be bounded?

The answer to this question lies in the kernels $k_2(x, \xi_1, \xi_2)$, $k_3(x, \xi_1, \xi_2, \xi_3)$, ..., their structure, and their rate of growth/decay.

14.3 PDEs for the Kernels of the Spatial Volterra Series in the Nonlinear Feedback Operator

Let us first note that the kernels of the Volterra series depend on a number of arguments which grow with the index of the kernel, namely,

$k_2(x, \xi_1, \xi_2)$	(3 arguments)
$k_3(x, \xi_1, \xi_2, \xi_3)$	(4 arguments)
$k_4(x, \xi_1, \xi_2, \xi_3, \xi_4)$	(5 arguments)
$k_5(x, \xi_1, \xi_2, \xi_3, \xi_4, \xi_5)$	(6 arguments)
$\vdots$	

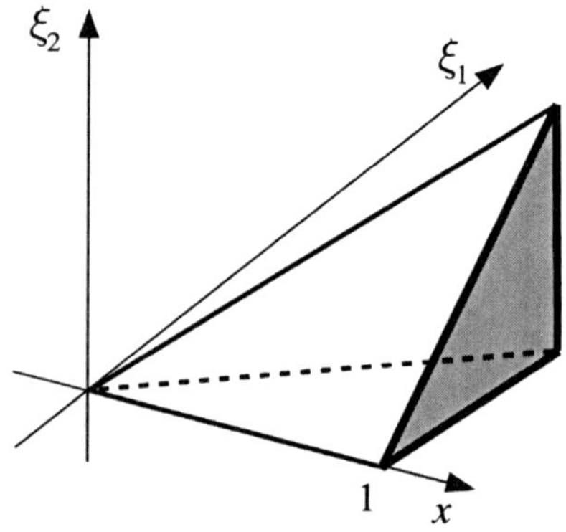

Figure 14.2. *The kernel PDE domain* $T_2 = \{(x, \xi_1, \xi_2) : 0 \le \xi_2 \le \xi_1 \le x \le 1\}$, *with the control kernel* $k_2(1, \xi_1, \xi_2)$ *defined in the shaded triangle.*

As was the case with the linear problems in the book so far, the kernels are governed by hyperbolic PDEs. The interesting fact, however, is that a *cascade* structure exists among the PDEs governing the sequence of kernels in the nonlinear case; namely, the PDE for the kernel k_2 is autonomous, the PDE for k_3 is governed by the solution for k_2, and so on:

$$k_2 \to k_3 \to k_4 \to k_5 \to \cdots \tag{14.15}$$

Next, by matching (14.7), (14.8) and (14.12), (14.13), with the help of the Volterra transformation (14.10), we obtain the following PDEs for the first two Volterra kernels. The PDEs for the subsequent kernels become increasingly more complicated, partly because of both the growing number of their arguments and the growing dimension of the PDEs. For this reason, we stop after showing the details of k_3.

PDE for kernel $k_2(x, \xi_1, \xi_2)$:

$$\partial_{xx} k_2 = \partial_{\xi_1\xi_1} k_2 + \partial_{\xi_2\xi_2} k_2 \qquad \text{(2D wave equation)}, \tag{14.16}$$

$$k_2(x, x, \xi_2) = \xi_2 - x\,, \tag{14.17}$$

$$\partial_x k_2(x, x, \xi_2) = -2\,, \tag{14.18}$$

$$\partial_{\xi_2} k_2(x, \xi_1, 0) = 0\,, \tag{14.19}$$

$$\partial_{\xi_1} k_2(x, \xi_1, \xi_1) = \partial_{\xi_2} k_2(x, \xi_1, \xi_1)\,. \tag{14.20}$$

The domain of this PDE is the pyramid

$$T_2 = \{(x, \xi_1, \xi_2) : 0 \le \xi_2 \le \xi_1 \le x \le 1\}\,, \tag{14.21}$$

shown in Figure 14.2, where the control kernel $k_2(1, \xi_1, \xi_2)$ is defined in the shaded triangle. Note that the kernel k_2 is nonzero, which is exclusively due to the inhomogeneities in the boundary conditions (14.17), (14.18).

PDE for kernel $k_3(x, \xi_1, \xi_2, \xi_3)$:

$$
\begin{aligned}
\partial_{xx} k_3 &= \partial_{\xi_1\xi_1} k_3 + \partial_{\xi_2\xi_2} k_3 + \partial_{\xi_3\xi_3} k_3 \\
&\quad + 4k_2(x, \xi_1, \xi_2) + 2k_2(x, \xi_1, \xi_3) \qquad \text{(3D wave equation)}, && (14.22)\\
k_3(x, x, \xi_2, \xi_3) &= 0\,, && (14.23)\\
\partial_x k_3(x, x, \xi_2, \xi_3) &= 3(x - \xi_2)^2\,, && (14.24)\\
\partial_{\xi_3} k_3(x, \xi_1, \xi_2, 0) &= 0\,, && (14.25)\\
\partial_{\xi_1} k_3(x, \xi_1, \xi_1, \xi_3) &= \partial_{\xi_2} k_3(x, \xi_1, \xi_1, \xi_3)\,, && (14.26)\\
\partial_{\xi_2} k_3(x, \xi_1, \xi_2, \xi_2) &= \partial_{\xi_2} k_3(x, \xi_1, \xi_2, \xi_2)\,. && (14.27)
\end{aligned}
$$

The domain of this PDE is the "hyperpyramid" $0 \le \xi_3 \le \xi_2 \le \xi_1 \le x \le 1$. Besides noting the inhomogeneity in boundary condition (14.22), one should also note that this PDE is driven by the solution for k_2.

14.4 Numerical Results

The numerical results for the example from the previous section are shown in Figures 14.3 and 14.4. The first term in the Volterra series for the nonlinear feedback law, $u(1) = \int_0^1 u(\xi_1) \int_0^{\xi_1} u(\xi_2) k_2(1, \xi_1, \xi_2) d\xi_2 d\xi_1$, is already successful in preventing the finite escape instability for an initial condition of significant size.

If one were interested in replacing the quadratic feedback law

$$u(1) = \int_0^1 u(\xi_1) \int_0^{\xi_1} u(\xi_2) k_2(1, \xi_1, \xi_2) d\xi_2 d\xi_1 \tag{14.28}$$

with a quadratic feedback law in the v-variable, one would obtain it as

$$v_x(1) = -v(1) \int_0^1 v(x) \partial_{\xi_2} k_2(1, 1, x) dx + \int_0^1 v(x) \int_0^x v(\xi) \partial_{\xi_1} \partial_{\xi_2} k_2(1, x, \xi) d\xi dx\,. \tag{14.29}$$

14.5 What Class of Nonlinear PDEs Can This Approach Be Applied to in General?

The nonlinear design approach presented in this chapter is applicable to the general class of plants *transformable* into the form

$$
\begin{aligned}
u_t = u_{xx} + \lambda(x) u &+ \sum_{n=1}^{\infty} \int_0^x \int_0^{\xi_1} \cdots \int_0^{\xi_{n-1}} f_n(x, \xi_1, \ldots, \xi_n) \left(\prod_{j=1}^{\infty} u(\xi_j, t) \right) d\xi_n \ldots \xi_1 \\
&+ u(x, t) \sum_{n=1}^{\infty} \int_0^x \int_0^{\xi_1} \cdots \int_0^{\xi_{n-1}} h_n(x, \xi_1, \ldots, \xi_n) \left(\prod_{j=1}^{\infty} u(\xi_j, t) \right) d\xi_n \ldots \xi_1\,,
\end{aligned} \tag{14.30}
$$

where the nonlinearities are given as Volterra series with kernels f_n and h_n.

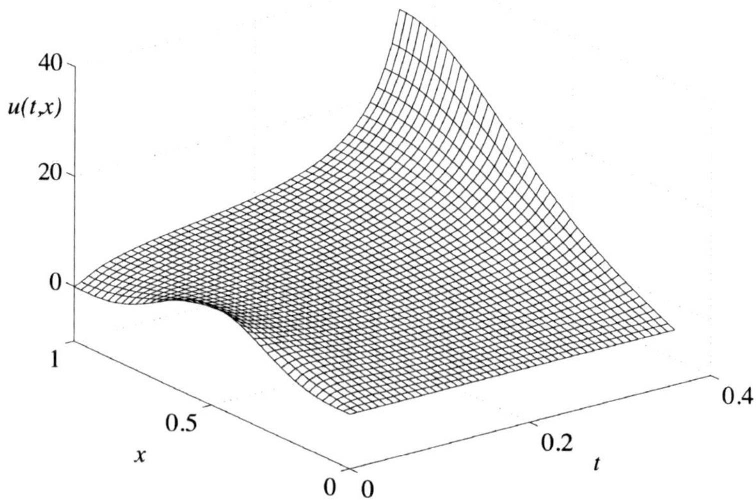

open-loop response for PDE $u_t(x,t) = u_{xx}(x,t) + 2u(x,t)\int_0^x u(\xi,t)d\xi$

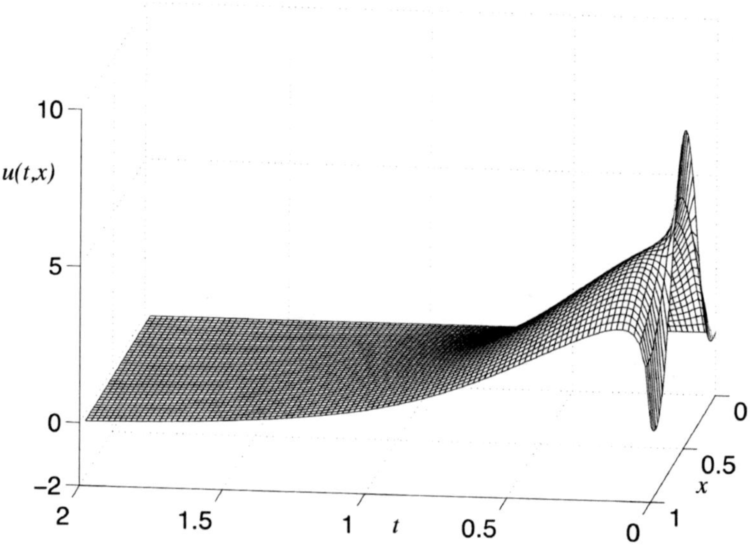

closed-loop with controller $u(1) = \int_0^1 u(\xi_1)\int_0^{\xi_1} u(\xi_2)k_2(1,\xi_1,\xi_2)d\xi_2 d\xi_1$

Figure 14.3. *Open-loop and closed-loop simulation results for the nonlinear PDE $u_t(x,t) = u_{xx}(x,t) + 2u(x,t)\int_0^x u(\xi,t)d\xi$. In open loop, the solution blows up in finite time. In closed loop, the PDE is stabilized to zero. Only the first significant term (the quadratic term) in the Volterra feedback is applied.*

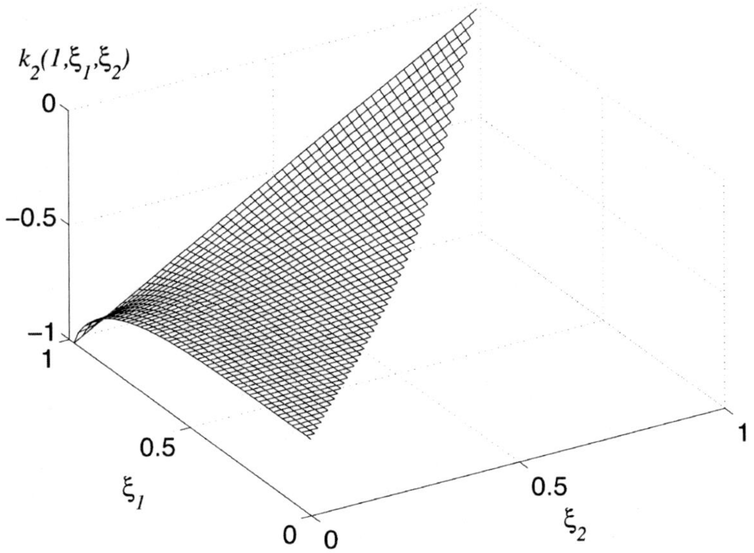

Figure 14.4. *The kernel of the quadratic term in the Volterra series feedback law.*

The motivation for this class of plants comes from applications. Consider the following system of coupled PDEs, which is a simplified example of dynamics that arise in chemical process control (reactant concentration dynamics coupled with thermal dynamics):

$$u_t = u_{xx} + \mu v\,, \tag{14.31}$$

$$\varepsilon v_t = v_{xx} + \omega^2 v + uv + u\,. \tag{14.32}$$

When the two subsystems are operating on very different time scales, namely, when $\varepsilon \to 0$, the mapping $u \mapsto v$ is a Volterra series in x (after an application of a boundary feedback $v(1)$ that eliminates a Fredholm operator from the solution for $v(x)$).

14.5.1 Some Bad News

We have observed, by examining the details of the design for a PDE with a simple reaction term, that the feedback synthesis is quite complex—it contains a Volterra series operator even when no such complex nonlinear operator is present in the plant. When the plant class is enlarged to (14.30), where Volterra series appear in the plant, one can expect a significant increase in the complexity of the feedback design. Indeed, the number of terms on the right-hand side of the k_n PDE in the design for the plant class (14.30) is a staggering $3 \cdot 2^n - n - 3$.

14.5.2 Some Good News

Fortunately, the *volume of the domain of evolution* of the kernel PDEs k_n decays as $\frac{1}{(n+1)!}$. Furthermore, the *volume of the domain of integration* of each term in Volterra series (for the transformation and the control law) decays as $\frac{1}{n!}$. As a result, the control law is convergent under very nonrestrictive assumptions on the plant nonlinearities, as detailed next.

14.5.3 What Can Actually Be Proved?

The following properties of the control law and of the feedback system can be proved [175, 176, 177]:

- **A priori (Lyapunov) estimates can be derived for each $k_n(x, \xi_1, \dots, \xi_n)$.** The growth of their L_2- and L_∞-norms in n can be quantified on domains of increasing dimension, i.e., for wave equations in nD, as $n \to \infty$.

- **The controller can be shown to be convergent.** For any plant nonlinearity with a globally convergent Volterra series, the Volterra series of the backstepping transformation and feedback law can be proved to be globally convergent for states in L_2 or L_∞. The meaning of this is that the plant Volterra kernels are not only allowed to be large, but they also are even permitted to have a certain rate of growth with n.

- **The closed-loop system is asymptotically stable.** This is a consequence of the stability of the target system and of the fact that the nonlinear backstepping transformation is *invertible*.

14.5.4 Is the Complexity Method Induced or Inherent?

The complexity in the design is inherent in the class of plants and the problem being pursued. The complexity may appear to be method induced when backstepping is considered only for a stabilization problem, where as many different solutions exist as control Lyapunov functions. However, the design procedure from this chapter applies also to solving a motion planning problem for the same class of systems. The motion planning problem is an "exact" problem, with a unique solution. Hence, the complexity observed in the solution to this problem is inherent in the problem, not method induced.

14.6 Notes and References

The material in this chapter is based on [175, 176, 177]. Numerous research opportunities exist for extending the nonlinear control designs for the reaction-diffusion partial integro-differential equations in this chapter to nonlinear versions of various other classes of PDE systems discussed in this book.

One regrettable feature of the control problems for nonlinear PDEs is that they are not always globally solvable. For nonlinear ODEs in the "strict-feedback" class, control problems (stabilization, motion planning, tracking, disturbance attenuation, adaptive control, etc.) are always solvable globally. Since this class is the ODE equivalent of the boundary control problems for nonlinear PDEs, one would hope for global solvability of similar problems for nonlinear PDEs. Unfortunately, this is not the case, as indicated through counter examples in [53, 63].

The failure of global solvability of stabilization (and motion planning) problems for nonlinear PDEs via the method of backstepping manifests itself through the lack of global invertibility of the backstepping transformation. While the direct backstepping transformation is always globally well defined, its inverse may be well defined only for a limited size of its functional argument.

Despite the disappointing observation that we may not be able to achieve global results for nonlinear PDE problems of interest, the backstepping design allows us to achieve stabilization in large regions of attraction, whose size can be estimated thanks to the availability of a Lyapunov function, which is constructed through the backstepping transformation.

Exercise

14.1. By matching (14.7), (14.8) and (14.12), (14.13), with the help of the Volterra transformation (14.10), derive the PDEs for the kernels k_2 and k_3 in (14.16)–(14.20) and (14.22)–(14.27).

Appendix

Bessel Functions

A.1 Bessel Function J_n

The function $y(x) = J_n(x)$ is a solution to the following ODE:

$$x^2 y''_{xx} + x y'_x + (x^2 - n^2) y = 0 . \tag{A.1}$$

Series representation:

$$J_n(x) = \sum_{m=0}^{\infty} \frac{(-1)^m (x/2)^{n+2m}}{m!(m+n)!} . \tag{A.2}$$

Properties:

$$2n J_n(x) = x(J_{n-1}(x) + J_{n+1}(x)) , \tag{A.3}$$

$$J_n(-x) = (-1)^n J_n(x) . \tag{A.4}$$

Differentiation:

$$\frac{d}{dx} J_n(x) = \frac{1}{2}(J_{n-1}(x) - J_{n+1}(x)) = \frac{n}{x} J_n(x) - J_{n+1}(x) , \tag{A.5}$$

$$\frac{d}{dx}(x^n J_n(x)) = x^n J_{n-1}, \qquad \frac{d}{dx}(x^{-n} J_n(x)) = -x^{-n} J_{n+1} . \tag{A.6}$$

Asymptotic properties (illustrated in Figure A.1):

$$J_n(x) \approx \frac{1}{n!}\left(\frac{x}{2}\right)^n, \qquad x \to 0 , \tag{A.7}$$

$$J_n(x) \approx \sqrt{\frac{2}{\pi x}} \cos\left(x - \frac{\pi n}{2} - \frac{\pi}{4}\right), \qquad x \to \infty . \tag{A.8}$$

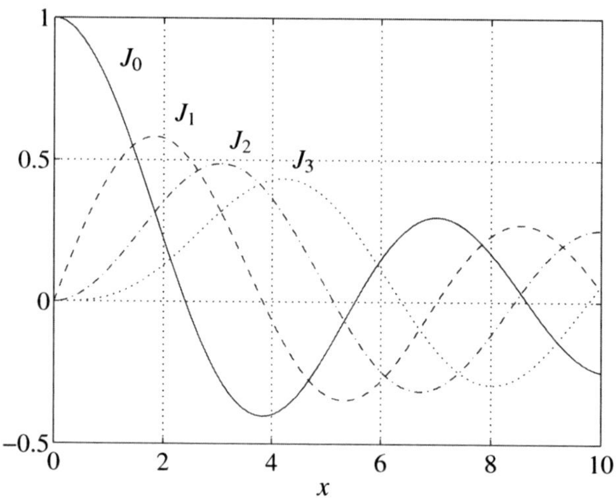

Figure A.1. *Bessel function J_n.*

A.2 Modified Bessel Function I_n

The function $y(x) = I_n(x)$ is a solution to the following ODE:

$$x^2 y''_{xx} + x y'_x - (x^2 + n^2) y = 0 . \tag{A.9}$$

Series representation:

$$I_n(x) = \sum_{m=0}^{\infty} \frac{(x/2)^{n+2m}}{m!(m+n)!} . \tag{A.10}$$

Relationship with $J_n(x)$:

$$I_n(x) = i^{-n} J_n(ix), \qquad I_n(ix) = i^n J_n(x) . \tag{A.11}$$

Properties:

$$2n I_n(x) = x(I_{n-1}(x) - I_{n+1}(x)) , \tag{A.12}$$

$$I_n(-x) = (-1)^n I_n(x) . \tag{A.13}$$

Differentiation:

$$\frac{d}{dx} I_n(x) = \frac{1}{2}(I_{n-1}(x) + I_{n+1}(x)) = \frac{n}{x} I_n(x) + I_{n+1}(x) , \tag{A.14}$$

$$\frac{d}{dx}(x^n I_n(x)) = x^n I_{n-1}, \qquad \frac{d}{dx}(x^{-n} I_n(x)) = x^{-n} I_{n+1} . \tag{A.15}$$

Asymptotic properties (illustrated in Figure A.2):

$$I_n(x) \approx \frac{1}{n!}\left(\frac{x}{2}\right)^n, \qquad x \to 0, \tag{A.16}$$

$$I_n(x) \approx \frac{e^x}{\sqrt{2\pi x}}, \qquad x \to \infty. \tag{A.17}$$

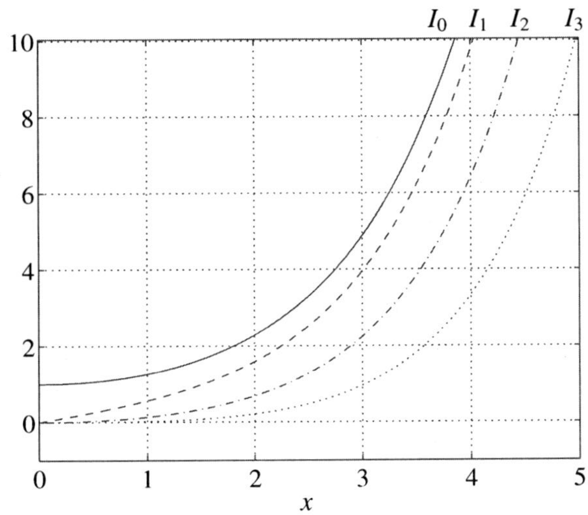

Figure A.2. *Modified Bessel function I_n.*

Bibliography

[1] O.-M. Aamo and M. Krstic, *Flow Control by Feedback: Stabilization and Mixing*, Springer, New York, 2003.

[2] ——, *Global stabilization of a nonlinear Ginzburg–Landau model of vortex shedding*, European Journal of Control, 10 (2004), pp. 105–118.

[3] O.-M. Aamo, A. Smyshlyaev, and M. Krstic, *Boundary control of the linearized Ginzburg–Landau model of vortex shedding*, SIAM Journal on Control and Optimization, 43 (2005), pp. 1953–1971.

[4] O.-M. Aamo, A. Smyshlyaev, M. Krstic, and B. Foss, *Stabilization of a Ginzburg–Landau model of vortex shedding by output-feedback boundary control*, IEEE Transactions on Automatic Control, 52 (2007), pp. 742–748.

[5] H. Amann, *Feedback stabilization of linear and semilinear parabolic systems*, in Semigroup Theory and Applications, Lecture Notes in Pure and Appl. Math., 116 (1989), Marcel Dekker, New York, pp. 21–57.

[6] Z. Artstein, *Linear systems with delayed controls: A reduction*, IEEE Transactions on Automatic Control, 27 (1982), pp. 869–879.

[7] J. Baker, A. Armaou, and P. D. Christofides, *Nonlinear control of incompressible fluid flow: Application to Burgers' equation and 2D channel flow,* Journal of Mathematical Analysis and Applications, 252 (2000), pp. 230–255.

[8] A. Balogh, O. M. Aamo, and M. Krstic, *Optimal mixing enhancement in 3D pipe flow*, IEEE Transactions on Control Systems Technology, 13 (2005), pp. 27–41.

[9] A. Balogh and M. Krstic, *Burgers' equation with nonlinear boundary feedback: H^1 stability, well posedness, and simulation*, Mathematical Problems in Engineering, 6 (2000), pp. 189–200.

[10] ——, *Boundary control of the Korteweg–de Vries–Burgers equation: Further results on stabilization and numerical demonstration*, IEEE Transactions on Automatic Control, 45 (2000), pp. 1739–1745.

[11] ——, *Infinite dimensional backstepping-style feedback transformations for a heat equation with an arbitrary level of instability*, European Journal of Control, 8 (2002), pp. 165–175.

[12] A. Balogh and M. Krstic, *Stability of partial differential equations governing control gains in infinite dimensional backstepping*, Systems and Control Letters, 51 (2004), pp. 151–164.

[13] A. Balogh, W.-J. Liu, and M. Krstic, *Stability enhancement by boundary control in 2D channel flow,* IEEE Transactions on Automatic Control, 46 (2001), pp. 1696–1711.

[14] B. Bamieh and M. A. Dahleh, *Energy amplification in channel flows with stochastic excitation*, Physics of Fluids, 13 (2001), pp. 3258–3269.

[15] B. Bamieh, F. Paganini, and M. A. Dahleh, *Distributed control of spatially-invariant systems*, IEEE Transactions on Automatic Control, 47 (2002), pp. 1091–1107.

[16] A. Banaszuk, H. A. Hauksson, and I. Mezić, *A backstepping controller for a nonlinear partial differential equation model of compression system instabilities*, SIAM Journal on Control and Optimization, 37 (1999), pp. 1503–1537.

[17] H. T. Banks, R. C. Smith, and Y. Wang, *Smart Material Structures: Modeling, Estimation, and Control*, Wiley, New York, Masson, Paris, 1996.

[18] V. Barbu, *Feedback stabilization of Navier–Stokes equations,* ESAIM: Control, Optimisation and Calculus of Variations, 9 (2003), pp. 197–205.

[19] V. Barbu, I. Lasiecka, and R. Triggiani, *Tangential boundary stabilization of Navier–Stokes equations,* Memoirs of the American Mathematical Society, 181 (2006), no. 852.

[20] V. Barbu and R. Triggiani, *Internal stabilization of Navier–Stokes equations with finite-dimensional controllers,* Indiana University Mathematics Journal, 53 (2004), pp. 1443–1494.

[21] J. Baumeister, W. Scondo, M. A. Demetriou, and I. G. Rosen, *On-line parameter estimation for infinite-dimensional dynamical systems*, SIAM Journal on Control Optimization, 35 (1997), pp. 678–713.

[22] A. Bensoussan, G. Da Prato, M. C. Delfour, and S. K. Mitter, *Representation and Control of Infinite-Dimensional Systems*, 2nd ed., Birkhäuser, Boston, 2006.

[23] J. Bentsman and Y. Orlov, *Reduced spatial order model reference adaptive control of spatially varying distributed parameter systems of parabolic and hyperbolic types*, International Journal of Adaptive Control and Signal Processing, 15 (2001), pp. 679–696.

[24] T. R. Bewley, *Flow control: New challenges for a new Renaissance*, Progress in Aerospace Sciences, 37 (2001), pp. 21–58.

[25] M. Böhm, M. A. Demetriou, S. Reich, and I. G. Rosen, *Model reference adaptive control of distributed parameter systems*, SIAM Journal on Control and Optimization, 36 (1998), pp. 33–81.

[26] D. Boskovic, A. Balogh, and M. Krstic, *Backstepping in infinite dimension for a class of parabolic distributed parameter systems*, Mathematics of Control, Signals, and Systems, 16 (2003), pp. 44–75.

[27] D. M. Boskovic and M. Krstic, *Backstepping control of chemical tubular reactor*, Computers and Chemical Engineering, 26 (2002), pp. 1077–1085.

[28] ——, *Stabilization of a solid propellant rocket instability by state feedback*, International Journal of Robust and Nonlinear Control, 13 (2003), pp. 483–495.

[29] ——, *Nonlinear stabilization of a thermal convection loop by state feedback*, Automatica, 37 (2001), pp. 2033–2040.

[30] D. M. Boskovic, M. Krstic, and W. J. Liu, *Boundary control of an unstable heat equation via measurement of domain-averaged temperature*, IEEE Transactions on Automatic Control, 46 (2001), pp. 2022–2028.

[31] J. A. Burns and K. P. Hulsing, *Numerical methods for approximating functional gains in LQR boundary control problems*, Mathematical and Computer Modeling, 33 (2001), pp. 89–100.

[32] J. A. Burns and D. Rubio, *A distributed parameter control approach to sensor location for optimal feedback control of thermal processes*, in Proceedings of the 36th Annual Conference on Decision and Control, San Diego, CA, Dec. 1997, pp. 2243–2247.

[33] H. S. Carslaw and J. C. Jaeger, *Conduction of Heat in Solids*, Clarendon Press, Oxford, 1959.

[34] ——, *Feedback control of hyperbolic PDE systems*, AIChE Journal, 42 (1996), pp. 3063–3086.

[35] P. D. Christofides and P. Daoutidis, *Robust control of hyperbolic PDE systems*, Chemical Engineering Science, 53 (1998), pp. 85–105.

[36] P. Christofides, *Nonlinear and Robust Control of Partial Differential Equation Systems: Methods and Applications to Transport-Reaction Processes*, Birkhäuser, Boston, 2001.

[37] J. Cochran, R. Vazquez, and M. Krstic, *Backstepping boundary control of Navier–Stokes channel flow: A 3D extension*, in Proceedings of the 25th American Control Conference (ACC), pp. 769–774, Minneapolis, 2006.

[38] D. Colton, *The solution of initial-boundary value problems for parabolic equations by the method of integral operators*, Journal of Differential Equations, 26 (1977), pp. 181–190.

[39] J.-M. Coron, *On the controllability of the 2D incompressible Navier–Stokes equations with the Navier slip boundary conditions*, ESAIM Control, Optimisation and Calculus of Variations, 1 (1996), pp. 35–75.

[40] R. Courant and D. Hilbert, *Methods of Mathematical Physics*, Interscience Publishers, New York, 1962.

[41] R. F. Curtain, *Robust stabilizability of normalized coprime factors: The infinite-dimensional case*, International Journal on Control, 51 (1990), pp. 1173–1190.

[42] R. F. Curtain, M. A. Demetriou, and K. Ito, *Adaptive observers for structurally perturbed infinite dimensional systems*, in Proceedings of the 36th Conference on Decision and Control, San Diego, CA, Dec. 1997, pp. 509–514.

[43] ——, *Adaptive observers for slowly time varying infinite dimensional systems*, in Proceedings of the 37th Conference on Decision and Control, Tampa, FL, Dec. 1998, pp. 4022–4027.

[44] R. F. Curtain and H. J. Zwart, *An Introduction to Infinite Dimensional Linear Systems Theory*, Springer-Verlag, New York, 1995.

[45] L. Debnath, *Nonlinear Partial Differential Equations for Scientists and Engineers*, Birkhäuser, Boston, 1997.

[46] M. A. Demetriou and K. Ito, *Optimal on-line parameter estimation for a class of infinite dimensional systems using Kalman filters*, in Proceedings of the American Control Conference, 2003, pp. 2708–2713.

[47] M. A. Demetriou and I. G. Rosen, *Variable structure model reference adaptive control of parabolic distributed parameter systems*, in Proceedings of the American Control Conference, Anchorage, AK, May 2002, pp. 4371–4376.

[48] ——, *On-line robust parameter identification for parabolic systems*, International Journal of Adaptive Control and Signal Processing, 15 (2001), pp. 615–631.

[49] D. G. Duffy, *Transform Methods for Solving Partial Differential Equations*, CRC Press, Boca Raton, FL, 1994.

[50] W. B. Dunbar, N. Petit, P. Rouchon, and P. Martin, *Motion planning for a nonlinear Stefan problem*, ESAIM Control, Optimisation and Calculus of Variations, 9 (2003), pp. 275–296.

[51] T. E. Duncan, B. Maslowski, and B. Pasik-Duncan, *Adaptive boundary and point control of linear stochastic distributed parameter systems*, SIAM Journal on Control and Optimization, 32 (1994), pp. 648–672.

[52] H. O. Fattorini, *Boundary control systems*, SIAM Journal on Control, 6 (1968), pp. 349–385.

[53] E. Fernandez-Cara and S. Guerrero, *Null controllability of the Burgers' equation with distributed controls*, Systems and Control Letters, 56 (2007), pp. 366–372.

[54] Y. A. Fiagbedzi and A. E. Pearson, *Feedback stabilization of linear autonomous time lag systems*, IEEE Transactions on Automatic Control, 31 (1986), pp. 847–855.

[55] M. Fliess and H. Mounier, *Tracking control and π-freeness of infinite dimensional linear system*, in Dynamical Systems, Control Coding, Computer Vision (G. Picci and D. S. Gilliam, eds.), Birkhäuser, Boston, 1999, pp. 45–68.

[56] M. Fliess, H. Mounier, P. Rouchon, and J. Rudolph, *A distributed parameter approach to the control of a tubular reactor: A multi-variable case*, in Proceedings of the 37th Conference on Decision and Control, Tampa, FL, 1998, pp. 736–741.

[57] S. S. Frazao and Y. Zech, *Undular bores and secondary waves—experiments and hybrid finite-volume modelling*, Journal of Hydraulic Research, 40 (2002), pp. 33–43.

[58] N. Fujii, *Feedback stabilization of distributed parameter systems by a functional observer*, SIAM Journal on Control and Optimization, 18 (1980), pp. 108–121.

[59] A. V. Fursikov, *Stabilization of two-dimensional Navier–Stokes equations with help of a boundary feedback control,* Journal of Mathematical Fluid Mechanics, 3 (2001), pp. 259–301.

[60] A. V. Fursikov, *Stabilization for the 3-D Navier–Stokes system by feedback boundary control,* Discrete and Continuous Dynamical Systems, 10 (2004), pp. 289–314.

[61] A. V. Fursikov, M. D. Gunzburger, and L. S. Hou, *Boundary value problems and optimal boundary control for the Navier–Stokes system: The two-dimensional case,* SIAM Journal on Control and Optimization, 36 (1998), pp. 852–894.

[62] K. Gu and S.-I. Niculescu, *Survey on recent results in the stability and control of time-delay systems*, Transactions of ASME, 125 (2003), pp. 158–165.

[63] S. Guerrero and O. Y. Imanuvilov, *Remarks on global controllability for the Burgers equation with two control forces*, Annales de l'Institut Henri Poincaré. Analyse Non Linéaire, 24 (2007), pp. 897–906.

[64] M. D. Gunzburger, *Perspectives in Flow Control and Optimization*, SIAM, Philadelphia, 2002.

[65] M. Gunzburger and H.C. Lee, *Feedback control of Karman vortex shedding*, Transactions of ASME, 63 (1996), pp. 828–835.

[66] M. D. Gunzburger and S. Manservisi, *The velocity tracking problem for Navier–Stokes flows with boundary control,* SIAM Journal on Control and Optimization, 39 (2000), pp. 594–634.

[67] B.-Z. Guo and Z.-C. Shao, *Regularity of a Schrödinger equation with Dirichlet control and colocated observation,* Systems and Control Letters, 54 (2005), pp. 1135–1142.

[68] S. M. Han, H. Benaroya, and T. Wei, *Dynamics of transversely vibrating beams using four engineering theories*, Journal of Sound and Vibration, 225 (1999), pp. 935–988.

[69] G. H. Hardy, J. E. Littlewood, and G. Polya, *Inequalities*, 2nd ed., Cambridge University Press, Cambridge, UK, 1959.

[70] J.-W. He, R. Glowinski, R. Metcalfe, A. Nordlander, and J. Periaux, *Active control and drag optimization for flow past a circular cylinder:* I. *Oscillatory cylinder rotation*, Journal of Computational Physics, 163 (2000), pp. 83–117.

[71] D. Henry, *Geometric Theory of Semilinear Parabolic Equations*, Springer, New York, 1993.

[72] M. Hogberg, T. R. Bewley, and D. S. Henningson, *Linear feedback control and estimation of transition in plane channel flow,* Journal of Fluid Mechanics, 481 (2003), pp. 149–175.

[73] K. S. Hong and J. Bentsman, *Direct adaptive control of parabolic systems: Algorithm synthesis and convergence and stability analysis*, IEEE Transactions on Automatic Control, 39 (1994), pp. 2018–2033.

[74] P. Huerre and P. A. Monkewitz, *Local and global instabilities in spatially developing flows*, Annual Review of Fluid Mechanics, 22 (1990), pp. 473–537.

[75] O. Y. Imanuvilov, *On exact controllability for the Navier–Stokes equations,* ESAIM Control, Optimisation and Calculus of Variations, 3 (1998), pp. 97–131.

[76] P. Ioannou and J. Sun, *Robust Adaptive Control*, Prentice–Hall, Upper Saddle River, NJ, 1996.

[77] M. Jankovic, *Control Lyapunov–Razumikhin functions and robust stabilization of time delay systems*, IEEE Transactions on Automatic Control, 46 (2001), pp. 1048–1060.

[78] M. R. Jovanovic and B. Bamieh, *Componentwise energy amplification in channel flows,* Journal of Fluid Mechanics, 543 (2005), pp. 145–183.

[79] ——, *Lyapunov-based distributed control of systems on lattices*, IEEE Transactions on Automatic Control, 50 (2005), pp. 422–433.

[80] I. Kanellakopoulos, P. V. Kokotovic, and A. S. Morse, *A toolkit for nonlinear feedback design*, Systems and Control Letters, 18 (1992), pp. 83–92.

[81] B. van Keulen, *H_∞-Control for Distributed Parameter Systems: A State-Space Approach*, Birkhäuser, Boston, 1993.

[82] H. Khalil, *Nonlinear Systems*, 2nd ed., Prentice–Hall, Upper Saddle River, NJ, 1996.

[83] J. U. Kim and Y. Renardy, *Boundary control of the Timoshenko beam*, SIAM Journal on Control and Optimization, 25 (1987), pp. 1417–1429.

[84] J. Klamka, *Observer for linear feedback control of systems with distributed delays in controls and outputs*, Systems and Control Letters, 1 (1982), pp. 326–331.

[85] T. Kobayashi, *Global adaptive stabilization of infinite-dimensional systems*, Systems and Control Letters, 9 (1987), pp. 215–223.

[86] T. Kobayashi, *Adaptive regulator design of a viscous Burgers' system by boundary control*, IMA Journal of Mathematical Control and Information, 18 (2001), pp. 427–437.

[87] ——, *Stabilization of infinite-dimensional second-order systems by adaptive PI-controllers*, Mathematical Methods in the Applied Sciences, 24 (2001), pp. 513–527.

[88] ——, *Adaptive stabilization of the Kuramoto–Sivashinsky equation*, International Journal of Systems Science, 33 (2002), pp. 175–180.

[89] ——, *Low-gain adaptive stabilization of infinite-dimensional second-order systems*, Journal of Mathematical Analysis and Applications, 275 (2002), pp. 835–849.

[90] ——, *Adaptive stabilization of infinite-dimensional semilinear second-order systems*, IMA Journal of Mathematical Control and Information, 20 (2003), pp. 137–152.

[91] V. Komornik, *Exact Controllability and Stabilization: The Multiplier Method*, Res. Appl. Math., 36, Wiley, Chichester, Masson, Paris, 1994.

[92] A. J. Krener and W. Kang, *Locally convergent nonlinear observers*, SIAM Journal of Control and Optimization, 42 (2003), pp. 155–177.

[93] M. Krstic, *On global stabilization of Burgers' equation by boundary control*, Systems and Control Letters, 37 (1999), pp. 123–142.

[94] ——, *Systematization of approaches to adaptive boundary stabilization of PDEs*, International Journal of Robust and Nonlinear Control, 16 (2006), pp. 812–818.

[95] M. Krstic and H. Deng, *Stabilization of Nonlinear Uncertain Systems*, Springer-Verlag, New York, 1998.

[96] M. Krstic, B.-Z. Guo, A. Balogh, and A. Smyshlyaev, *Control of a tip-force destabilized shear beam by observer-based boundary feedback*, SIAM Journal on Control and Optimization, 47 (2008), pp. 553–574.

[97] M. Krstic, I. Kanellakopoulos, and P. Kokotovic, *Nonlinear and Adaptive Control Design*, Wiley, New York, 1995.

[98] M. Krstic and A. Smyshlyaev, *Adaptive boundary control for unstable parabolic PDEs—Part* I: *Lyapunov design*, IEEE Transactions on Automatic Control, to appear.

[99] M. Krstic, A. Siranosian, and A. Smyshlyaev, *Backstepping boundary controllers and observers for the slender Timoshenko beam: Part* I*—Design*, in Proceedings of the American Control Conference, 2006, pp. 2412–2417.

[100] M. Krstic, A. Siranosian, A. Smyshlyaev, and M. Bement, *Backstepping boundary controllers and observers for the slender Timoshenko beam: Part II—Stability and simulations*, in Proceedings of the 45th IEEE Conference on Decision and Control, San Diego, 2006, pp. 3938–3943.

[101] N. Kunimatsu and H. Sano, *Stability analysis of heat-exchanger equations with boundary feedbacks*, IMA Journal of Mathematical Control and Information, 15 (1998), pp. 317–330.

[102] W. H. Kwon and A. E. Pearson, *Feedback stabilization of linear systems with delayed control*, IEEE Transactions on Automatic Control, 25 (1980), pp. 266–269.

[103] J. E. Lagnese, *Boundary Stabilization of Thin Plates*, SIAM, Philadelphia, 1989.

[104] B. Laroche and P. Martin, *Motion planning for 1-D linear partial differential equations*, in Algebraic Methods in Flatness, Signal Processing and State Estimation, (H. Sira-Ramirez, G. Silva-Navarro, eds.), Mexico, 2003, pp. 55–76.

[105] B. Laroche, P. Martin, and P. Rouchon, *Motion planning for the heat equation*, International Journal of Robust and Nonlinear Control, 10 (2000), pp. 629–643.

[106] I. Lasiecka, *Mathematical Control Theory of Coupled PDEs*, SIAM, Philadelphia, 2002.

[107] ——, *Stabilization and structural assignment of Dirichlet boundary feedback parabolic equations*, SIAM Journal on Control and Optimization, 21 (1983), pp. 766–803.

[108] ——, *Optimal regularity, exact controllability and uniform stabilisation of Schrödinger equations with Dirichlet control,* Differential Integral Equations, 5 (1992), pp. 521–535.

[109] I. Lasiecka and R. Triggiani, *Control Theory for Partial Differential Equations: Continuous and Approximation Theories*, 2 vols., Cambridge University Press, Cambridge, UK, 2000.

[110] E. Lauga and T. R. Bewley, *Performance of a linear robust control strategy on a nonlinear model of spatially-developing flows*, Journal of Fluid Mechanics, 512 (2004), pp. 343–374.

[111] ——, *Modern control of linear global instability in a cylinder wake model*, International Journal of Heat and Fluid Flow, 23 (2002), pp. 671–677.

[112] X. Li and J. Yong, *Optimal Control Theory for Infinite Dimensional Systems*, Birkhäuser, Boston, 1995.

[113] W. J. Liu, *Boundary feedback stabilization of an unstable heat equation*, SIAM Journal on Control and Optimization, 42 (2003), pp. 1033–1043.

[114] W.-J. Liu and M. Krstic, *Backstepping boundary control of Burgers' equation with actuator dynamics*, Systems and Control Letters, 41 (2000), pp. 291–303.

[115] W.-J. Liu and M. Krstic, *Stability enhancement by boundary control in the Kuramoto–Sivashinsky equation,* Nonlinear Analysis, 43 (2000), pp. 485–583.

[116] ——, *Adaptive control of Burgers' equation with unknown viscosity,* International Journal of Adaptive Control and Signal Proccessing, 15 (2001), pp. 745–766.

[117] ——, *Global boundary stabilization of the Korteweg–de Vries–Burgers equation,* Computational and Applied Mathematics, 21 (2002), pp. 315–354.

[118] H. Logemann, *Stabilization and regulation of infinite-dimensional systems using coprime factorizations,* in Analysis and Optimization of Systems: State and Frequency Domain Approaches for Infinite-Dimensional Systems (Sophia-Antipolis, 1992), Lecture Notes in Control and Inform. Sci., 185, Springer, Berlin, 1993, pp. 102–139.

[119] H. Logemann and B. Martensson, *Adaptive stabilization of infinite-dimensional systems,* IEEE Transactions on Automatic Control, 37 (1992), pp. 1869–1883.

[120] H. Logemann and E. P. Ryan, *Time-varying and adaptive integral control of infinite-dimensional regular linear systems with input nonlinearities,* SIAM Journal of Control and Optimization, 38 (2000), pp. 1120–1144.

[121] H. Logemann and S. Townley, *Adaptive stabilization without identification for distributed parameter systems: An overview,* IMA Journal of Mathematical Control and Information, 14 (1997), pp. 175–206.

[122] Z. H. Luo, B. Z. Guo, and O. Morgul, *Stability and Stabilization of Infinite Dimensional Systems with Applications,* Springer-Verlag, New York, 1999.

[123] A. Macchelli and C. Melchiorri, *Modeling and control of the Timoshenko beam. The distributed port Hamiltonian approach,* SIAM Journal of Control and Optimization, 43 (2004), pp. 743–767.

[124] E. Machtyngier, *Exact controllability for the Schrödinger equation,* SIAM Journal on Control and Optimization, 32 (1994), pp. 24–34.

[125] E. Machtyngier and E. Zuazua, *Stabilization of the Schrödinger equation,* Portugaliae Mathematica, 51 (1994), pp. 243–256.

[126] A. Z. Manitius and A. W. Olbrot, *Finite spectrum assignment for systems with delays,* IEEE Transactions on Automatic Control, 24 (1979), pp. 541–553.

[127] F. Mazenc and P.-A. Bliman, *Backstepping designs for time-delay nonlinear systems,* IEEE Transactions on Automatic Control, 51 (2006), pp. 149–154.

[128] T. Meurer and M. Zeitz, *Flatness-based feedback control of diffusion-convection-reactions systems via k-summable power series,* in Preprints of NOLCOS 2004, Stuttgart, Germany, 2004, pp. 191–196.

[129] S. Mondie and W. Michiels, *Finite spectrum assignment of unstable time-delay systems with a safe implementation*, IEEE Transactions on Automatic Control, 48 (2003), pp. 2207–2212.

[130] O. Morgul, *Dynamic boundary control of the Timoshenko beam*, Automatica, 28 (1992), pp. 1255–1260.

[131] T. Nambu, *On the stabilization of diffusion equations: Boundary observation and feedback*, Journal of Differential Equations, 52 (1984), pp. 204–233.

[132] A. W. Naylor and G. R. Sell, *Linear Operator Theory in Engineering and Science*, Springer-Verlag, New York, 1982.

[133] A. W. Olbrot, *Stabilizability, detectability, and spectrum assignment for linear autonomous systems with general time delays*, IEEE Transactions on Automatic Control, 23 (1978), pp. 887–890.

[134] F. Ollivier and A. Sedoglavic, *A generalization of flatness to nonlinear systems of partial differential equations. Application to the command of a flexible rod*, in Proceedings of the 5th IFAC Symposium on Nonlinear Control Systems, vol. 1, 2001, pp. 196–200.

[135] Y. Orlov, *Sliding mode observer-based synthesis of state derivative-free model reference adaptive control of distributed parameter systems*, Journal of Dynamic Systems, Measurements, and Control, 122 (2000), pp. 726–731.

[136] Y. Orlov and J. Bentsman, *Adaptive distributed parameter systems identification with enforceable identifiability conditions and reduced-order spatial differentiation*, IEEE Transactions Automatic Control, 45 (2000), pp. 203–216.

[137] D. S. Park, D. M. Ladd, and E. W. Hendricks, *Feedback control of von Kármán vortex shedding behind a circular cylinder at low Reynolds numbers*, Physics of Fluids, 6 (1994), pp. 2390–2405.

[138] N. Petit and D. Del Vecchio, *Boundary control for an industrial under-actuated tubular chemical reactor*, Journal of Process Control, 15 (2005), pp. 771–784.

[139] K.-D. Phung, *Observability and control of Schrödinger equations,* SIAM Journal of Control and Optimization, 40 (2001), pp. 211–230.

[140] A. D. Polianin, *Handbook of Linear Partial Differential Equations for Engineers and Scientists*, Chapman and Hall/CRC, Boca Raton, FL, 2002.

[141] L. Praly, *Adaptive regulation: Lyapunov design with a growth condition*, International Journal of Adaptive Control and Signal Processing, 6 (1992), pp. 329–351.

[142] B. Protas and A. Styczek, *Optimal control of the cylinder wake in the laminar regime,* Physics of Fluids, 14 (2002), pp. 2073–2087.

[143] A. P. Prudnikov, Yu. A. Brychkov, and O. I. Marichev, *Integrals and Series, Vol.* 2: *Special Functions*, Gordon and Breach, New York, 1986.

[144] M. S. DE QUEIROZ, D. M. DAWSON, M. AGARWAL, AND F. ZHANG, *Adaptive nonlinear boundary control of a flexible link robot arm*, IEEE Transactions on Robotics and Automation, 15 (1999), pp. 779–787.

[145] M. S. DE QUEIROZ, D. M. DAWSON, S. P. NAGARKATTI, AND F. ZHANG, *Lyapunov-Based Control of Mechanical Systems*, Birkhäuser, Basel, 2000.

[146] J.-P. RAYMOND, *Feedback boundary stabilization of the two dimensional Navier–Stokes equations,* SIAM Journal on Control and Optimization, 45 (2006), pp. 790–828.

[147] P. ROUCHON, *Motion planning, equivalence, infinite dimensional systems*, International Journal of Applied Mathematics and Computer Science, 11 (2001), pp. 165–188.

[148] K. ROUSSOPOULOS AND P. A. MONKEWITZ, *Nonlinear modelling of vortex shedding control in cylinder wakes*, Physica D, 97 (1996), pp. 264–273.

[149] D. L. RUSSELL, *Differential-delay equations as canonical forms for controlled hyperbolic systems with applications to spectral assignment*, in Control Theory of Systems Governed by Partial Differential Equations, A. K. Aziz, J. W. Wingate, M. J. Balas, eds., Academic Press, New York, 1977, pp. 119–150.

[150] ——, *Controllability and stabilizability theory for linear partial differential equations: Recent progress and open questions*, SIAM Review, 20 (1978), pp. 639–739.

[151] H. SANO, *Exponential stability of a mono-tubular heat exchanger equation with output feedback*, Systems and Control Letters, 50 (2003), pp. 363–369.

[152] T. I. SEIDMAN, *Two results on exact boundary control of parabolic equations*, Applied Mathematics and Optimization, 11 (1984), pp. 145–152.

[153] D.-H. SHI, S. H. HOU, AND D.-X. FENG, *Feedback stabilization of a Timoshenko beam with an end mass*, International Journal of Control, 69 (1998), pp. 285–300.

[154] A. SMYSHLYAEV AND M. KRSTIC, *Closed form boundary state feedbacks for a class of 1D partial integro-differential equations*, IEEE Transactions on Automatic Control, 49 (2004), pp. 2185–2202.

[155] ——, *Backstepping observers for a class of parabolic PDEs*, Systems and Control Letters, 54 (2005), pp. 613–625.

[156] ——, *On control design for PDEs with space-dependent diffusivity or time-dependent reactivity*, Automatica, 41 (2005), pp. 1601–1608.

[157] ——, *Lyapunov adaptive boundary control for parabolic PDEs with spatially varying coefficients*, in Proceedings of the 25th American Control Conference, Minneapolis, 2006, pp. 41–48.

[158] ——, *Adaptive boundary control for unstable parabolic PDEs—Part* II: *Estimation-based designs*, Automatica, 43 (2007), pp. 1543–1556.

[159] A. Smyshlyaev and M. Krstic, *Adaptive boundary control for unstable parabolic PDEs—Part* III: *Output feedback examples with swapping identifiers*, Automatica, 43 (2007), pp. 1557–1564.

[160] ——, *Adaptive Schemes for Boundary Control of Parabolic PDEs*, in preparation.

[161] V. Solo and B. Bamieh, *Adaptive distributed control of a parabolic system with spatially varying parameters*, in Proceedings of the 38th IEEE Conference on Decision and Control, 1999, pp. 2892–2895.

[162] S. S. Sritharan, *Stochastic Navier–Stokes equations: Solvability, control and filtering,* in Stochastic Partial Differential Equations and Applications – VII, G. Da Prato and L. Tubaro eds., Chapman & Hall/CRC, Boca Raton, FL, 2006, pp. 273–280.

[163] R. Temam, *Navier–Stokes Equations*, Studies in Mathematics and Its Applications, Vol. 2, North-Holland Publishing Co., Amsterdam, 1984.

[164] ——, *Infinite-Dimensional Dynamical Systems in Mechanics and Physics*, Springer, New York, 1988.

[165] J.-P. Thibault and L. Rossi, *Electromagnetic flow control: Characteristic numbers and flow regimes of a wall-normal actuator,* Journal of Physics D: Applied Physics, 36 (2003), pp. 2559–2568.

[166] A. N. Tikhonov and A. A. Samarskii, *Equations of Mathematical Physics. Sixth Supplemental and Revised Edition,* 2 vols., Edwin Mellen Press, New York, 1999.

[167] S. Townley, *Simple adaptive stabilization of output feedback stabilizable distributed parameter systems*, Dynamics and Control, 5 (1995), pp. 107–123.

[168] R. Triggiani, *Well-posedness and regularity of boundary feedback parabolic systems,* Journal of Differential Equations, 36 (1980), pp. 347–362.

[169] ——, *Boundary feedback stabilization of parabolic equations*, Applied Mathematics and Optimization, 6 (1980), pp. 201–220.

[170] ——, *Exponential feedback stabilization of a 2-D linearized Navier–Stokes channel flow by finite-dimensional, wall-normal boundary controllers, with arbitrarily small support,* preprint, 2006.

[171] R. Vazquez and M. Krstic, *A closed form feedback controller for stabilization of linearized Navier–Stokes equations: The 2D Poisseuille flow*, in Proceedings of the IEEE Conference on Decision and Control, 2005, pp. 7358–7365.

[172] ——, *A closed form observer for the channel flow Navier–Stokes system*, in Proceedings of the IEEE Conference on Decision and Control, 2005, pp. 5959–5964.

[173] ——, *Higher order stability properties of a 2D Navier–Stokes system with an explicit boundary controller*, in Proceedings of the 25th American Control Conference, Minneapolis, 2006, pp. 1167–1172.

[174] R. Vazquez and M. Krstic, *Explicit integral operator feedback for local stabilization of nonlinear thermal convection loop PDEs*, Systems and Control Letters, 55 (2006), pp. 624–632.

[175] ——, *Boundary control laws for parabolic PDEs with Volterra nonlinearities – Part* I: *Design*, in Proceedings of the 7th IFAC Symposium on Nonlinear Control Systems, Pretoria, South Africa, 2007.

[176] ——, *Boundary control laws for parabolic PDEs with Volterra nonlinearities – Part* II: *Examples*, in Proceedings of the 7th IFAC Symposium on Nonlinear Control Systems, Pretoria, South Africa, 2007.

[177] ——, *Boundary control laws for parabolic PDEs with Volterra nonlinearities – Part* III: *Analysis*, in Proceedings of the 7th IFAC Symposium on Nonlinear Control Systems, Pretoria, South Africa, 2007.

[178] R. Vazquez and M. Krstic, *Control of Turbulent and Magnetohydrodynamic Channel Flows*, Birkhäuser, Boston, 2008.

[179] R. Vazquez, E. Schuster, and M. Krstic, *A closed-form observer for the 3D inductionless MHD and Navier–Stokes channel flow,* in Proceedings of the 45th IEEE Conference on Decision and Control, San Diego, 2006, pp. 739–746.

[180] R. Vazquez, E. Trélat, and J.-M. Coron, *Stable Poiseuille flow transfer for a Navier–Stokes system,* in Proceedings of the American Control Conference, Minneapolis, 2006, pp. 775–780.

[181] R. B. Vinter and R. H. Kwong, *The infinite time quadratic control problem for linear systems with state and control delays: An evolution approach*, SIAM Journal on Control and Optimization, 19 (1981), pp. 139–153.

[182] J. A. Walker, *Dynamical Systems and Evolution Equations*, Plenum, New York, 1980.

[183] K. Watanabe and M. Ito, *An observer for linear feedback control laws of multivariable systems with multiple delays in controls and outputs*, Systems and Control Letters, 1 (1981), pp. 54–59.

[184] J. T.-Y. Wen and M. J. Balas, *Robust adaptive control in Hilbert space*, Journal of Mathematical Analysis and Applications, 143 (1989), pp. 1–26.

[185] X. Yu and K. Liu, *Eventual regularity of the semigroup associated with the monotubular heat exchanger equation with output feedback*, Systems and Control Letters, 55 (2006), pp. 859–862.

[186] H. L. Zhao, K. S. Liu, and C. G. Zhang, *Stability for the Timoshenko beam system with local Kelvin–Voigt damping*, Acta Mathematica Sinica, English Series, 21 (2005), pp. 655–666.

[187] E. ZAUDERER, *Partial Differential Equations of Applied Mathematics*, 2nd ed., Wiley, New York, 1998.

[188] F. ZHANG, D. M. DAWSON, M. S. DE QUEIROZ, AND P. VEDAGARBHA, *Boundary control of the Timoshenko beam with free-end mass/inertial dynamics*, in Proceedings of the 36th IEEE Conference on Decision and Control, Vol. 1, 1997, pp. 245–250.

Index